SHATTERING
the MYTHS *of*
DARWINISM

SHATTERING
the MYTHS *of*
DARWINISM

RICHARD MILTON

Park Street Press
Rochester, Vermont

Park Street Press
One Park Street
Rochester, Vermont 05767
www.InnerTraditions.com

Park Street Press is a division of Inner Traditions International

Copyright © 1992, 1997 by Richard Milton
First U.S. edition published by Park Street Press in 1997

Originally published in Great Britain by Fourth Estate Limited

LIBRARY OF CONGRESS CATALOGING-IN-PUBLICATION DATA
    Milton, Richard, 1943–
        [Facts of life]
        Shattering the myths of Darwinism / Richard Milton.
            p.   cm.
        "Originally published in Great Britain by Fourth Estate Limited"—
    T.p. verso.
        Originally published: Facts of life : London : Corgi Books, 1993.
        Includes bibliographical references (p.   ) and indexes.
        ISBN 0-89281-732-1 (alk. paper)
        ISBN 0-89281-884-0 (paperback)
        1. Evolution (Biology)—Philosophy. I. Title.
    QH360.5.M55   1997
    576.8'2—DC21   97-9962
                                                                            CIP

Printed and bound in the United States

10 9 8 7 6 5 4 3 2 1

Text design and layout by Kristin Camp
This book was typeset in Janson with Lucida Sans as the display typeface

For Julia

Contents

AFTERWORD: CONTROVERSIES

Preface

WHEN THE FIRST EDITION OF THIS BOOK was published, in 1992, it was greeted with a storm of controversy no less fervent than the debate that met the publication of Darwin's theory of evolution one hundred and thirty years ago.

On one hand, according to a leading article in *The Times*; "Richard Milton's *Shattering the Myths of Darwinism* . . . could shake the 'religion' of evolution as much as *Honest to God* shook popular Christianity 30 years ago."[1]

While on the other, according to a review by Darwinist Richard Dawkins, the book is "loony," "stupid," "drivel" and its author a "harmless fruitcake" who "needs psychiatric help."[2]

When *Shattering the Myths of Darwinism* was published, I expected it to arouse controversy, because it reports on scientific research that is itself controversial and because it deals with Darwinism—always a touchy subject with the biology establishment.

I didn't expect science to welcome an inquisitive reporter, but I did expect the controversy to be conducted at a rational level, that people would rightly demand to inspect my evidence more closely and question me on the correctness of this or that fact. To my horror, I found that instead of challenging me, orthodox scientists simply set about seeing me off "their" property.

Richard Dawkins, a reader in zoology at Oxford University, wrote his review for the *New Statesman* magazine "lest the paper commission someone else who would treat it as a serious scientific treatise."[3] Dawkins devoted two-thirds of his review to attacking my British publishers, Fourth Estate, for their irresponsibility in daring to accept a book criticizing Darwinism and the remainder to assassinating my own character in the sort of terms quoted above.

Dawkins is employed at one of Britain's most distinguished universities and is responsible for the education of future generations of students. Yet this is not the language of a responsible scientist and teacher. It is the language of a religious fundamentalist whose faith has been profaned.

Nature magazine, probably the most highly respected scientific magazine in the world, scented blood and joined in the frenzy. Its editor, John Maddox, ran a leading article that described me as believing science to be a myth (I don't), evolution to be false (I don't), and natural selection to be a pack of lies (I don't).[4] It also magisterially rebuked the *Sunday Times* for daring to devote most of one of its main news pages to reporting the book's disclosures.[5]

These intemperate responses betoken more than a squabble between an inquisitive journalist and a couple of reactionary academics. They raise a number of important questions of general public interest. Who do you have to be to have a voice about scientific research on which large sums of public money are spent? Who *decides* who you have to be? In what forum, or by what mechanism, can the voices of dissent ever be heard in science?

It is not just outsiders who cannot be heard, it is dissenting members of the scientific professions themselves. In my mailbox are letters from biologists who are concerned by the teaching of Darwinism as holy writ and from medical men whose practices have led them to make medical discoveries having a bearing on evolutionary biology. They have sought to publicize these discoveries in journals such as *Nature* but have been universally rejected because their discoveries are anti-Darwinian in implication and hence counter to the ruling ideology in the life sciences. They have appealed to me—a nonscientist—to help them gain publicity.

It is worrying to learn that in countries like Britain and the United States, thought to be among the most civilized on Earth, some professional scientists can feel so isolated and ignored that they have to take their case directly to the public via the popular press. Equally, it is depressing to discover that in countries which pride themselves on their intellectual tolerance, it is impossible to voice scientific dissent without attracting this kind of response from those who perceive themselves to be the guardians of orthodoxy.

In seeking to defend the ideological citadel of Darwinism, the most vociferous critics of this book have allowed their emotions to mislead them so far as to attack me for advocating beliefs that I have never held and do not support. Both Richard Dawkins and *Nature* have tried to suggest that I do not believe in evolution and that I believe the Earth is merely a few thousand years old.

To forestall any repetition of false claims like these, let me make my position clear on both issues from the outset. I accept that there is persuasive circumstantial evidence for evolution, but I do not accept that there is *any* significant evidence that the mechanism driving that evolution is the neo-Darwinian mechanism of chance mutation coupled with natural selection. Second, I do not believe that the Earth is only a few thousand years old. I present evidence that currently accepted methods of dating are seriously flawed and are supported by Darwinists only because they provide the billions of years required by Darwinist theories. Because radioactive dating methods are scientifically unreliable, it is at present impossible to say with any confidence how old the Earth is.

PART ONE

CHAOS

CHAPTER 1

A National Treasure

B Y THE RIVER THAMES AT TEDDINGTON, west of London, straggles a cluster of nondescript factory buildings that seem an unlikely home for a national treasure. Yet the buildings are of national importance to everyone in the United Kingdom because of a few pieces of metal kept there.

The buildings are those of the National Physical Laboratory, and the pieces of metal are platinum standards kept at strictly regulated temperature in air-conditioned chambers to act as the indisputable authority for Britain's national weights and measures, the pound and the kilogram, the yard and the meter.

Paradoxically, these standards are never used for everyday purposes. No one goes to Teddington to measure out a meter of cloth or a foot of parcel string. But their mere existence—unchanging and unchallengeable—is the nation's guarantee that standards exist: that should it ever be necessary, it is theoretically possible to make a physical comparison with the accepted measure and say with absolute certainty that the subject under test either is or is not of the stated weight or length.

It's not easy for members of the public to visit the National Physical Laboratory, because it is a busy government research establishment and the repository of many secrets. But it is certainly possible. Those fortunate enough to be given access to the NPL will see the famous platinum kilogram standard and the atomic clock that is the authority for Greenwich Mean Time. You can check your wristwatch and, in principle at least, check up on your 12-inch ruler and the weights of your kitchen scales.

A few miles to the east of Teddington stands the very much more imposing structure of the British Museum of Natural History, famous to generations of schoolchildren for its dinosaurs and dramatic reconstructions of the Earth's geological history. This building, too, is the repository of a "national standard" but one that is not on display in a glass case and that has proved very much more difficult to track down. The museum is one of the world's primary sources or authorities for the theory of evolution by natural selection, the theory that is taught in schools and universities the world over: a kind of headquarters for Darwinism.

Like millions of people, I have visited the museum many times to stare in wonder at its contents. But I have been unable to see with my own eyes the decisive evidence for the general theory of evolution. I have been able to see many marvels and to study mountains of evidence: the Geological Column that reconstructs the geological and biological history of the Earth; the dinosaur skeletons and myriad other fossils; marvels like the skeleton of *Archaeopteryx*, seemingly half bird, half reptile; the reconstructed evolution of the horse family. But unlike its counterpart at Teddington, the museum is unable to exhibit the unchallengeable authority that conclusively demonstrates that evolution by natural selection has taken place and is established as fact.

This is very far from saying that scientists have failed to make the case for Darwinian evolution. On the contrary, no rational person can visit this or any other Natural History Museum and not be deeply impressed by the evidence that has been painstakingly assembled: evidence of historical development over geological time, of similarity of anatomical structure in many different species, of change and adaptation to changing environments. But, frustratingly, even with all this evidence, it is impossible for the genuinely objective person to say, "Here is the conclusive scientific proof that I have been looking for."

My disappointment among the glass cabinets of Kensington was the beginning of a ten-year journey to try to find that conclusive proof. It has been an Odyssey that has taken me far both geographically and in time. As a science reporter I am used to beating a path from library to museum to laboratory and back again. Now I had to become a scientific historian as well and even a scientific detective to find the evidence I needed to examine afresh. That evidence is of many different kinds and is found in incongruous and often strange settings: in remote quarries and coastal cliffs, in libraries and museums, and even in bank vaults. Some can be seen only with the aid of the electron microscope and some has proved not to exist at all.

But although this book has taken me a decade or more to research, my immediate reason for writing it is a simple personal dilemma. My daughter Julia has just celebrated her ninth birthday. She is quickly developing an interest in natural history and like many nine-year-olds is an avid fossil collector and dinosaur-spotter. She is beginning to have serious science lessons at her school and is just now being introduced to the idea that life on Earth has arisen spontaneously from a common ancestor in the remote past and that all the species of animal and plant alive today have evolved from earlier species. Over the next ten years, she will be taught that the mechanism governing this process is that of genetic mutation and natural selection—the neo-Darwinist or synthetic theory of evolution.

The fact that Julia is beginning to absorb the general theory of evolution has been giving me sleepless nights. Is Julia being taught the truth? Or is she—and are we—being seriously misled?

Let me make it clear that my doubts about the general theory of evolution do not arise from religious objections. I want my daughter to have access to the fruits of scientific enquiry, whatever those findings should prove to be. But I am seriously concerned, on purely rational grounds, that generations of school and university teachers have been led to accept speculation as scientific theory and faulty data as scientific fact; that this process has accumulated a mountainous catalog of mingled fact and fiction that can no longer be contained by the sparsely elegant theory; and that it is high time that the theory was taken out of its ornate Victorian glass cabinet and examined with a fresh and skeptical eye.

My doubt about the theory arises from a number of sources. It comes first and most importantly from the inability of Darwinists to pass the simple test described earlier; to show a thinking member of the public conclusive scientific evidence to substantiate the theory, in the same way that the National Physical Laboratory can demonstrate physical constants, the College of Surgeons can demonstrate the circulation of the blood, or the Greenwich Observatory can demonstrate the expanding universe.

Second, it comes from the world of scientific investigation itself, a world that I write about in my job as a science reporter and where many discoveries that have an important bearing on evolution theory have been made in the last two decades, but have received little publicity.

Third, and perhaps most eloquently, it comes from the disarmingly direct questions of a nine-year-old child: Where did we come from? How old is the Earth? Do the butterfly and the elephant really have a common ancestor?

Today, the neo-Darwinist or synthetic theory of evolution enjoys unrivalled prominence as the only rational theory to account for the origin of species and the evolution of all creatures including humankind. It is the only theory of evolution taught in schools, colleges, and universities. It is presented as fact in museum displays, lectures, and publications. A few controversial points are referred to in museum publications and biology textbooks, but these are viewed as peripheral controversies, whose outcome cannot alter the basic truth revealed by neo-Darwinism. The synthetic theory is universally taken as having been confirmed in all its main essentials—only a few isolated details remaining to be tidied up by specialists in esoteric disciplines such as molecular biology.

The teachings of this theory are familiar to everyone educated in a Western country in the past fifty years. The Earth is of immense antiquity, formed 4,600 million years ago; life on this planet is also of very great age—emerging spontaneously in ancient seas 3,000 million years ago—and the great variety of species that exist today are all descended from one or a few primitive organisms evolved in those ancient oceans, by a process of random genetic mutation coupled with natural selection. These ideas are the cornerstones of modern historical geology and of our contemporary world view, as familiar to the elementary school pupil as to the postgraduate student of biology or geology.

To most students, teachers, and even some scientists it will come as a surprise to learn that recent research into the age of the Earth has produced evidence that our planet could be much younger than had previously been thought: existing methods of geochronometry such as uranium-lead decay and radiocarbon assay have been found to be deeply flawed and unreliable; the extent of genetic change by selection has been found experimentally to be limited; bacteria can be induced in the laboratory to mutate in a direction that is beneficial to them—without generations of natural selection; only a catastrophist model of development can account for important Earth structures and processes such as continental drift and most fossil-bearing rock formations—most of the Earth's surface in fact. These major discoveries have had profound consequences for the neo-Darwinist theory of evolution, yet few of them have found their way into the public domain, still less into school or university textbooks or museum displays.

This book attempts to make accessible, and put into context, these new discoveries to enable nonscientists to evaluate for themselves the status of the general theory of evolution and the new light cast on existing theories by the latest discoveries.

A number of books attacking the neo-Darwinist theory—and evolution in general—have been published in recent decades by religious writers seeking to promulgate biblical creation as an alternative viewpoint, and I should make it clear at the outset that this book is not in any sense a contribution to creationist literature, nor does it represent the creationist viewpoint (although some creationist objections to neo-Darwinism that have a basis in scientific research are included here).

Because some of the scientific matters discussed in this book are highly controversial, I have included references to original studies wherever necessary. However, responsibility for the conclusions drawn from these sources is mine alone.

CHAPTER 2

Through the Looking Glass

W HEN A FAIR-SKINNED WOMAN LIES ON A SUNNY BEACH for any length of time, her skin will acquire a tan in response to the Sun's ultraviolet rays. The longer she lies on the beach, the darker her tan will become. But no matter how long she lies on the beach, her children will never be born with her suntan.

This everyday experience is at the very heart of the approach to evolution taken by Darwin and his successors. Although apparently simple, it encapsulates an immensely sophisticated and detailed body of reasoning and natural observations. But it also embodies a fundamental belief that runs counter to intuition, that has proved impossible to confirm experimentally, and that Darwinists and their opponents have fought tooth and nail over for more than 130 years.

Darwinists believe that the reason it is impossible for a child to inherit characteristics its parents acquire during their lifetime (such as the mother's suntan) is that evolution is not supervised by any directing force or design, but by chance genetic mutation operating together with natural selection. In the example given, this means that the black and brown people of the world have not acquired dark pigmentation because they live in sunny regions or because it would be useful to them to have dark skin to screen the sun's rays. Instead, chance alone has given one or a few ancestors darker skin and natural selection favored the survival of the darker-skinned people because they lived in sunny regions.

Possibly a darker-skinned person was born in northern latitudes from time to time, but his or her skin gave no special survival advantage and any offspring did not prosper especially so the predominant skin tone remains white.

Modern Darwinists have actually taken this idea one step further and added a codicil to the theory from the field of molecular biology. The reason that acquired characteristics cannot be inherited, many believe, is because the mechanism of inheritance—the genes contained in our sexual cells—cannot be constructively affected by the environment. The genetic code is a one-way system. Information can be read out when a new life is generated but information cannot be written in to alter the characteristics of that new life.[1] If the offspring differs at all radically from its forebears, believe Darwinists, it is because of random chance and nothing more. The fundamental mechanism of evolution is, in Professor Jacques Monod's memorable phrase, "chance and necessity."[2] In the Darwinian world it is possible to look into the mirror of genetics and even to read what is written there but, unlike Alice, we can never go through the looking glass.

The Darwinian idea is powerful and beautiful. It has sustained evolution theory for a century and a half. Many discoveries after Darwin's death have tended to confirm the idea. And many new ideas have been advanced concerning, for instance, the origin of life from nonliving materials that make use of the Darwinian concept and that stand head and shoulders above any competing theory.[3]

And yet many people, both scientists and laymen, have entertained nagging doubts. Do we really believe that black people are black by accident? What kind of accident was it? Why don't we see such accidents happening today? Why does the fossil record not show us such accidents happening in the past? Once the questions begin, it is difficult to know where to stop.

For instance, if we don't see genetic mutations—the accidents of inheritance—happening because they are very rare, then how can there have been enough of them to produce anything as complex as humans? Darwinists say this is because many billions of years have elapsed since the Earth cooled. The geological strata covering the Earth's surface have taken millions of years to lay down and the fossil creatures in them lived millions of years ago. But if these rocks take millions of years to form, why do we find trees forty feet tall in the vertical position of growth in coal seams?[4] And if nature can produce such rich diversity as the present animal and plant kingdoms by pure chance, why is it that thousands of years of serious guided selection by humans has resulted only in trivial sub-specific variation of domestic plants and animals, while not one new species has been created?

These and hundreds of similar questions begin to press for answers. But before the Darwinian response to those questions can be evaluated, it is necessary first to gain a firm grasp of just exactly what modern Darwinists do believe and, perhaps more importantly, how they came to believe it.

The story begins appropriately enough in a garden—not that of Eden but rather the Botanical Gardens in Ghent where, in 1898, a 27-year-old Austrian botanist called Erich Tschermak began to take an interest in breeding garden peas. After only two years' work, trying to breed distinctive characteristics, he found to his astonishment that the hybrids showed a mathematically precise ratio of yellow-seeded to green-seeded peas. Reading the literature on peas he found a cross-reference that seemed interesting and in 1900 he sent to the library of Vienna University for the papers.

At Amsterdam University, the Professor of Botany, Hugo de Vries, made a similarly exciting discovery in 1886. He found certain wild varieties of the evening primrose that differed markedly from the cultivated variety. Coining the term "mutation" to describe the phenomenon, de Vries started a long series of plant breeding experiments to see if he could breed mutations. In 1900 he made a breakthrough that started him searching through the University's literature on pea breeding.

In 1892, the instructor of botany at the University of Tubingen, Germany, Carl Erich Correns, conducted some research that required him to breed garden peas, and in 1900 he found the same mathematical pattern that his two contemporaries had discerned in their experiments. He, too, searched the literature in the university library and like his colleagues was astonished to discover that the entire subject had been researched and published in great detail a generation before by an unknown Augustinian monk, Gregor Mendel.

The simultaneous independent discovery by the three men—
all of whom later had distinguished careers in biology—came at a
time when Darwin's theory had been all but consigned to the scrap
heap of history. The theory had tottered on for a few years following Darwin's death but had fallen out of favor because it lacked a
credible mechanism that could cause change to take place in the
species that populate the world. Darwin had suggested natural fluctuations in form, gently edging a species in one direction rather
than another: like the giraffe's neck getting imperceptibly longer
with each generation. But this phenomenon was nowhere observed
in nature. Stability is the norm, not change—however slow—and
Darwin's idea of "heritable characters" was simply a nonstarter.

When Hugo de Vries found his markedly changed evening
primroses in the wild, he immediately surmised that he had found
such a mechanism: not the trivial "fluctuations" of Darwin but substantial "mutations" that accounted for bigger, sudden changes in
form. These mutations must be caused by radical changes in the
basic program of heredity—what were later called genes. What de
Vries and his codiscoverers had observed in pea plants was the tendency of certain genetic characteristics to dominate in a majority
of offspring—for most of them to be tall when a short pea was
crossed with a tall one. And it was this property, today called the
first law of genetics, that Gregor Mendel had discovered with his
pioneering efforts in the monastery garden in the 1860s.

It has often been remarked as strange that Mendel's great
achievement was never realized or acknowledged in his lifetime.
His paper was circulated to all of Europe's great libraries and was
received by many eminent biologists (including Darwin), none of
whom saw the importance of his discoveries. But if the European
mind was unreceptive in the 1860s, it had become highly receptive
four decades later, and Darwinian evolution theory was rescued
from the scrap heap at a single stroke. Combined with Mendelian
genetics, and the concept of mutation, Darwinism reemerged with
a solid experimental foundation as the neo-Darwinist or synthetic
theory of evolution.

In the neo-Darwinist theory, species evolve into other forms
by means of natural selection as Darwin had suggested. But they
do so not because of the trivial variation that occurs between all

individuals but because of chance mutations in their genetic makeup, most of which are neutral or lethal, but a few of which favor a change to a more advantageous form. Thus blind chance combines with necessity to shape the animal and plant kingdoms.

From the rediscovery of Mendel until today, the synthetic theory has remained preeminent as the only scientific theory to account for the origin of species and the evolution of all creatures including mankind. No other theory is taught in secondary schools (with the possible exception of the few religious schools that remain outside the state system) or in universities and colleges.

In the United Kingdom, for instance, the new National Curriculum for schools lays down these instructions for teachers of biology: "Pupils should develop their knowledge and understanding of variation and its genetic and environmental causes and the basic mechanisms of inheritance, selection and evolution." The National Curriculum's Attainment Target on "Genetics and Evolution" specifies pupils' objective as, "Understanding the relationship between variation, natural selection and reproductive success in organisms and the significance of their relationship for evolution."

So strong has the Darwinian model of evolution become that it has vanquished and displaced once and for all the lingering challengers it once had: Lamarckism or the inheritance of acquired characteristics; various versions of vitalism—the idea that evolution is supervised by some nonphysical natural force; and of course the biblical account of creation by an almighty hand. It is unsurprising that the principal opponents of evolution theory, both in its original form as conceived by Darwin and in its modern synthetic form, have been individuals who accept the biblical account of creation, who see Darwinism representing an assault on their religious beliefs, and who wish to have the theory given much reduced prominence, especially in education.

The movement has been active for most of this century, especially in the United States. From the Scopes trial in Tennessee in 1925 to creationist pressure groups of the 1990s, religious fundamentalists in America have repeatedly challenged evolutionists whatever their hue, sometimes successfully, sometimes not.[5] Some creationists are also scientists—a few of considerable academic standing—and they have offered substantive scientific criticisms of neo-Darwinism (some of which are included in this book).

An important factor in bringing about the universal dominance and acceptance of Darwinian evolution has been that virtually every eminent professional scientist appointed to posts in the life sciences in the last 40 or 50 years, in the English-speaking world, has been a convinced Darwinist. These eminent names include men such as Gavin de Beer (Professor of Embryology at University College London 1945–50 and Director of the British Museum of Natural History 1950–60), Julian Huxley (Professor of Zoology at King's College, London University, and Secretary to the London Zoological Society), J. B. S. Haldane (Professor of Genetics at London University 1933–37 and Professor of Biometry at University College 1937–57), and C. H. Waddington (Professor of Biology at Edinburgh University). In the United States, leading synthetic evolutionists have included Ernst Mayr (Professor of Zoology at Harvard University 1953–61 and Director of the Museum of Comparative Zoology 1961–70), Theodosius Dobzhansky (Professor of Zoology at Columbia University 1942–60), and George Simpson (Professor of Paleontology at Columbia University, Professor of Paleontology at the Harvard Museum of comparative zoology 1958–68, Professor of Geosciences at University of Arizona).

One of the most spirited of Darwin's champions from the non-English-speaking world has been France's Nobel prize–winning molecular biologist and Director of the Pasteur Institute Jacques Monod, whose 1970 book *Chance and Necessity* caused something of a shock wave on both sides of the Atlantic for its uncompromising portrayal of life as no more than chemistry and statistics.

These men, as well as occupying powerful and leading academic teaching positions, were also prolific and important writers whose influence has been widespread in forming the consensus. In Britain, Darwin's theory has been almost a family business for the Huxleys: Thomas Huxley acting as Darwin's champion and grandson Julian becoming an equally eminent biologist. Julian Huxley's *Evolution, the Modern Synthesis*, revised in 1963, is probably the closest published work to a complete all-embracing textbook on synthetic evolution (and is a valuable starting point for any enquirer on the subject).

In setting out to criticize neo-Darwinism, I have made these writers my chief sources of authority for articulating the modern synthetic theory in order not to misrepresent the current neo-Darwinist position. I have quoted primarily from their works (as well as from the writings of some of their close colleagues and adherents) so that the beliefs and ideas described are accurately representative rather than merely easy targets for refutation.

To dissent from the dominant scientific idea of the life sciences in the twentieth century may seem both foolhardy and unnecessary. After all, how could so many important scientists be wrong? Surely only religious cranks question evolution—the earnest kind who want to sell you strange newspapers and eagerly seek conversations "about life and its meaning"?

My reason for setting out to reevaluate the received wisdom of synthetic evolution is that something almost the opposite of these sentiments is actually the case. Far from being the province of cranks, it is the non-Darwinian view that is supported by modern findings. A very few (albeit very able and distinguished) scientists were responsible for making the synthetic theory preeminent in the natural sciences. They were able to do this in an era when intellectual authority often counted as much as experimental accuracy or innovation. And the principal findings that undermine the Darwinian theory have come from a new generation of scientists, less concerned with authoritarian theories and more concerned with unravelling mysteries.

Their findings have arisen from research in every one of the complex interlocking set of disciplines that go to make up the Darwinian theory: geology, stratigraphy, petrology, radioactive dating, paleontology, comparative anatomy, biology, zoology, genetics, molecular biology, organic chemistry. These findings undermine and challenge many fundamental tenets on which the theory is constructed; tenets as elementary as the age of the Earth, the formation of sedimentary rocks and the formation of the main features of the Earth's crust, the limits to specific variation, the causes of extinctions, and even the possible origins of life—long considered settled in broad outline. Yet these new findings have been given short shrift by the ruling ideology of science, primarily because of the huge investment of time, money, resources, and scientific reputations that neo-Darwinism represents.

But of course, there is much more to science's commitment to neo-Darwinism than careerism. It is an elegant, comprehensive, rationally based and economical set of neatly interlocking ideas that provide an important basis for understanding one of the most mysterious areas of scientific study—the origin and development of life.

I said earlier that I intend to take the theory out of the glass cabinet in which it is so reverently kept and to look at it a little less reverently and a little more closely. I want to begin this examination with a closer look at what is probably the central issue: the age of the Earth. The reason that this issue assumes key importance is because the central mechanism of neo-Darwinism, genetic mutation, means that change has to take place at an agonizingly slow pace—requiring hundreds of millions or even billions of years. If the Earth is of such an age, then neo-Darwinism could be true. If the Earth is not of such an age, then the theory cannot be true—despite what other evidence may indicate.

CHAPTER 3

A Matter of Conjecture

S O, HOW OLD IS THE EARTH? On the face of it, the answer to this question is cut-and-dried and no longer relevant to a discussion of evolution. It seems irrelevant to the modern debate because the age now universally accepted for the Earth is so vast—4,600 million years—as to allow life to have evolved not once but many times. But let us use our imaginations for a moment to ask two heretical questions. Does an age of 4,600 million years really provide enough time for evolution to have worked along Darwinian lines? And—even more outrageous—what if the Earth is not as old as we think?

Try this thought experiment on the first question. What has to happen for life to get started in the primeval oceans, and to develop by mutation and natural selection into the animal and plant kingdoms we see today? First, the inert chemicals in the sea must form amino acids, probably under the influence of ultraviolet light and electrical discharges in the form of lightning. This process was demonstrated by Harold Urey and Stanley Miller at Chicago University in 1953. In step two, the first amino acids in the early ocean must combine to form the stuff of life, protein molecules. It is these giant and complex molecules that ultimately constitute all plant and animal life, but the mechanism by which they might have formed spontaneously is not known and has not been demonstrated in the laboratory.

The Darwinian view is that although the formation of protein molecules without any precursor is highly improbable, it could have occurred given enough time—hundreds of millions of years. The third step will be the explosive variation and growth of all manner of life forms based on protein, from bacteria to Beethoven, again requiring hundreds of millions of years. Given steps one and two, this is not impossible to imagine, and from a Darwinian standpoint it would perhaps be surprising if it did not happen.

It is steps two and three that have a bearing on the age of the Earth. Although step two, the spontaneous formation of protein molecules, is an unknown process, it is theoretically possible to assess how long it would take to happen by chance. On the basis of size and complexity of such molecules, Murray Eden, Professor of Electrical Engineering at MIT, calculated that a very simple synthesis would be expected to happen by chance once in about 1,000 million years.[1] On the face of it, even these very lengthy odds can easily be accommodated in the 4,600 million years that most geologists assign to the Earth's history. But look a little closer.

Life is not spontaneously emerging today in the seas. This is attributed by Darwinists to the fact that conditions have changed since life evolved in the archaic oceans.[2] So how long exactly were the conditions suitable for this spontaneous happening? The time available would be bounded by two events. The cooling of the Earth and the establishing of the oceans would be the earlier marker event. This is said to have occurred about 3,800 million years ago (the date when the oldest known sedimentary rocks were formed).[3]

The upper marker would be the date of the first fossil of a living thing. Just where this upper marker should be placed is a controversial matter. The conservative view is that the first sign of life is represented by organisms called *Eobacterium isolatum* and *Archaeospheroides barbertonensis*, which are dated from about 3,200 million years ago.[4] This gives us a window of opportunity for the spontaneous occurrence of the first microorganism of roughly 600 million years. Actually, the gap is smaller than suggested by this crude sum because it would take considerable time for the new oceans to acquire the right mixture of basic chemicals to make the primeval "soup," and at the other end, bacteria must have been predated by some simpler nonreplicating molecules of which no trace survives. But let us be generous and allow the full 600 million years. What is a few million years when we have so many at our disposal?

This interval must accommodate not only the spontaneous combination of basic materials into amino acids, but also the combination of amino acids into protein molecules, the appearance of at least one self-replicating molecule and the subsequent evolution of this molecule into self-replicating cellular bodies to the bacterial level. And remember that of these four steps, one alone (the second) has been estimated to happen by chance once in 1,000 million years.

So, of the 4,600 million years of geological time that Darwinists have granted themselves, only a small fraction—less than 600 million—is actually available to accommodate the processes they believe to have taken place. Darwinian evolutionary processes are already running short of time.

The latest fossil evidence indicates that the gap is much narrower and that Darwinists have run out of time completely. Studying the oldest known sedimentary rocks from Greenland, said to date from 3,800 million years ago, geologists Hans Pflug and H. Jaeschke-Boyer found fossil cell-like structures in 1979 which they named *Isosphaera*.[5] The fossils are those of a primitive yeastlike organism. In 1980, C. Walters, A. Shimoyama, and C. Ponnamperuma examined *Isosphaera* for evidence of photosynthetic activity and announced ". . . we have now what we believe is strong evidence for life on Earth 3,800 million years ago."[6]

These findings were supported in 1988 by Manfred Schidlowski of the Max Planck Institute for Chemistry in Germany, who published a paper in *Nature* interpreting the proportion of light carbon isotopes in the 3,800-million-year-old sedimentary rocks from Isua in Greenland as signs of early organic life.[7] Schidlowski's interpretation was confirmed in 1996 by Gustaf Arrhenius of the Scripps Institute in San Diego who examined 3,800-million-year-old rocks from Isua and reported a mixture of carbon isotopes that only living things could produce.[8]

The meaning of these discoveries is unambiguously clear. If the first surface water formed 3,800 million years ago and the first microorganisms came into existence 3,800 million years ago, then there was zero time available for the spontaneous appearance of life. Life, it seems, did not wait for blind chance to roll the dice, but erupted at the first available instant, leaving Darwinists with no time at all for their probabilistic processes.

Strictly speaking, Darwinism is not concerned with abiogenesis—the appearance of life from inanimate matter—but only with the subsequent evolution of those primitive organisms into more highly developed species. In practice, however, Darwinism is intimately related to theories of abiogenesis. Darwin himself famously speculated in private correspondence about life coming into being spontaneously in some primitive warm pool. More significantly, all the plausible theories of abiogenesis that have so far been suggested employ the Darwinian mechanism of variation and natural selection—theories like that of Graham Cairns-Smith of Glasgow University who suggested that life arose by employing clays as catalysts.[9]

The discovery that such hypothetical processes had negligible time in which to bring about the first protein molecules and the first self-replicating organisms by chance is significant in light of the work of information theory scientist Hubert Yockey, who calculated the probability that a protein containing one hundred amino acids would form spontaneously as only 1 chance in 10^{65} at best.[10]

In 1989 Robert Sauer and his biologist colleagues at the Massachusetts Institute of Technology experimented with "re-building" proteins by taking away amino acids and replacing them with other amino acids. They found that some parts of a protein chain are tolerant to substitutions but other parts are completely intolerant of such tinkering, showing that proteins are not arbitrary collections of component chemicals but rare and sometimes unique combinations. Sauer and his colleagues confirmed Yockey's calculations that the probability of a specific folded protein coming into being by undirected evolution is 1 in 10^{65}. The practically infinite number of other combinations that could form at random are useless protein sequences that perform no constructive function for living organisms.[11]

These findings indicate that the magnitude of the improbability of proteins and self-replicating DNA molecules forming by chance is so great as to be virtually impossible in the time we now know was available. The probability calculated by Yockey and confirmed by Sauer's experiments—1 chance in 10^{65}—is an event so improbable that it could be compared to winning the state lottery by finding the winning ticket in the street, and then continuing to win the lottery every week for a thousand years, finding the winning ticket in the street each time—possible, in principle, if you have eternity at your disposal, but impossible, in practice, if all you have is a negligibly short time.

Darwinists are not in the slightest dismayed by such improbabilities for they can always fall back on the claim that however improbable the accidents needed for the first protein molecules to come into being, they must have come about, or else we would not be here.

A proper consideration of the role of probability in Darwinian theory must wait until a later chapter. For now, consider a second searching question; one that is even more heretical. Let us ask what evidence we have for the age of the Earth and what grounds we have for accepting that evidence.

The importance of this question, as observed earlier, lies in the fact that an Earth of immense age is indispensably necessary to the neo-Darwinist theory because genetic mutation and natural selection are processes that are conceived of as working very slowly over hundreds of millions of years. If the Earth were only a few million years old then there simply would not have been enough time for natural selection to work. Whether we liked it or not, we would be compelled to seek a fresh explanation for the origin of living species.

On this fundamentally important question, the Natural History Museum and all other modern authorities are in complete agreement. The Earth is 4,600 million years old. What is more, different periods of the Earth's history have been characterized by the formation of different kinds of rock containing the fossil remains of distinctive kinds of creature. These different periods have also been dated to give what is usually referred to as the Geological Column of the Earth's history. The column is reproduced on page 69.

By referring to the geological column anyone can tell the age of a rock or fossil that he or she finds. For instance, England's white cliffs consist of chalk dating from the end of the Cretaceous period, which, the column tells us, dates from 65 million years ago.

The dates attached to the geological column have been arrived at and refined over the past century or so. The most recent evaluation, and the one quoted in Natural History Museum publications, is that of Van Eysinga published in 1975.[12] This scheme (which is the source of Figure 1) is closely similar to that used in most museums and universities since the early decades of this century, and is based on the pioneering work of Arthur Holmes in the United Kingdom and Henry Faul in the United States. Some minor disagreements may exist among geologists but a very wide measure of agreement exists over the big issue that the earliest rocks of the column are around 4 billion years old, and over most of the details in Figure 1, for example that the Cretaceous period began around 140 million years ago and ended around 65 million years ago.

When I began to research this question a little more closely I uncovered a puzzle. Those experts I referred to and the authoritative textbooks I consulted all told me that modern dating has been accomplished by using radioactive methods and hence was an *absolute* dating method of a far higher order of accuracy than all previous methods—most of which relied on calculations involving one or more relative factors. These relative dating methods had relied on such factors as the increasing salinity of the oceans, or the Earth's rate of cooling, and are now considered unreliable. Radioactive dating, though, is used to date the rocks and the fossils they contain directly and hence was welcomed as an absolute method.

The puzzle arises because radioactive dating techniques can be applied only to volcanic rocks that contain some radioactive mineral—the primary rocks of the Earth's crust. But the geological column consists of sedimentary rocks—rocks formed from sediments laid down on the beds of ancient seas and composed of particles of those primary rocks. So, of course, any age determination made using these particles will be the same as that of the primary rocks from which they were derived. In some common sedimentary rocks, such as chalk or limestone, there are not even particles of the primary rocks present and so radioactive dating cannot be used at all. Happily for English men and women, the white cliffs of Dover are not radioactive.

In *The Age of the Earth* published by the Institute of Geological Sciences, the position is succinctly explained by John Thackray:

> The only sediments which can be dated directly are those in which a radioactive mineral is formed during diagenesis [laying down] of the sediment, such as the rather uncommon illite shales and glauconitic sandstones; other sediments give only the age of the parent rock from which the mineral grains that make them up are derived.[13]

How then did Holmes, Faul, and Van Eysinga arrive at the dates attached to the sediments of the geological column?

The Institute of Geological Sciences explains:

> Where lavas or volcanic ashes are interbedded with a sediment of known stratigraphic age, then a date may be given to that stratigraphic division. Where an igneous rock intrudes one sedimentary unit and is blanketed by another, then the sediments may be dated from the igneous rock by inference. The rarity of such cases, together with analytical error inherent in age determination, mean that isotopic ages are unlikely to rival or replace fossils as the most important means of . . . correlation.

It turns out that what has been dated by radioactive decay methods is not the sedimentary rocks or fossils themselves but the isolated intrusion into them of igneous or primary rocks, usually as volcanic material. This has been a rare and purely fortuitous process and one that is unreliable—so rare and so unreliable that the Institute of Geological Sciences thinks it unlikely to replace or even rival fossils as a method of dating. Nor is this all, for the method depends in turn on a further chain of inference. For the geological column of Van Eysinga is nowhere to be found in nature. It is an imaginary structure that has been synthesized from comparing a stratum of rock in one part of the world with a similar looking stratum in another part of the world (see chapter 7 for a more detailed discussion of the composition of the geological column).

Naturalists themselves are often confused in their knowledge of this question. Gavin de Beer, for example, director of the British Museum of Natural History from 1950 to 1960, wrote in the introduction to the museum's guide to evolution, published in 1970, that the rocks forming the geological column had been dated by radioactive methods:

> Estimates of time based on disintegration of radioactive material enable various levels of evolutionary lineages to be dated and the time measured during which certain changes have occurred, thereby providing quantitative evidence of evolution rates and the duration times of genera and species.[14]

This claim, which is universally believed and taught in schools and universities throughout the world, is entirely false. And when Darwinists speak of absolute dating of the geological column and the fossils it contains by radioactive methods they are quite mistaken, there is nothing absolute about it. In fact the method ought to be referred to as "comparative dating," because it dates the sedimentary rocks by inference alone through their relationship to the rare samples of igneous or primary rocks that are being dated.

When I pursued this question a little further, I found that there is in reality another factor that has been used to arrive at the age of the geological column and the fossils it contains —conjecture. This process crept into geological dating at a very early stage when Charles Lyell, the nineteenth century's most prominent geologist and Darwin's mentor in geological matters, attempted to date the end of the Cretaceous period by reference to how long he thought it would have taken the shellfish (whose fossils are found in later beds) to have evolved into their modern descendants. Lyell estimated that the Cretaceous ended 80 million years ago—not too far from today's accepted figure of 65 million, plus or minus 3 million.

According to Harold Levin of Washington University, "By comparing the amount of evolution exhibited by the marine molluscs in the various series of the Tertiary System with the amount that had occurred since the beginning of the Pleistocene Ice Age, Lyell estimated that 80 million years had elapsed since the beginning of the Cenozoic."[15]

Levin adds that, "He came astonishingly close to the mark." In fact, it is not at all astonishing when you know that today's accepted date has been derived not from an absolute, independent source but from conjectures including Lyell's.

The kind of surmise used to supplement the relative dates yielded by radioactive dating includes assumptions about the rates at which sediments are laid down on the bottoms of lakes, seashores, and ocean floors; estimates of the rates at which forests are turned into coal deposits; and estimates of the rates at which certain very long-lived families of creatures might have evolved. But although these conjectures are embodied in the modern view of the age of geological deposits, they are rarely if ever disclosed in geological or biological textbooks, and they are rarely exposed to debate.

Curiously, too, no geologist seems to have checked out the geological column dates with an electronic calculator on a common-sense basis. Let us go back to the illustration of the column in Figure 1 and look again at the thickness of the rocks in each period compared with the length of time assigned to those periods. Note that there is a remarkable consistency between assigned age and thickness of deposit. For instance the Cretaceous period is said to have lasted 65 million years and is 15,000 meters thick—an average annual rate of deposition of 0.2 millimeters. Now look at the Silurian period: this, too, yields an average rate of deposition of about 0.2 millimeters per year—as does the Ordovician, the Devonian, the Carboniferous, and the rest. It is only when we come to relatively modern times in the Cenozoic era that rates of deposition vary much, and here they appear to speed up slightly.

This is a very remarkable finding. One naturally expects Uniformitarian geology to favor uniformity, but this is too much of a good thing. Throughout widely changing climatic conditions, advancing and retreating oceans, droughts, and Ice Ages, the rate of sedimentation appears to remain amazingly constant regardless—throughout the thousands of millions of years that are said to have elapsed. The presumed rate of deposition itself—about the thickness of a human hair in a year—is a matter looked at in more detail later. But it is worth pausing in passing to note that such a slow rate would be quite incapable of burying and fossilizing entire forests, dinosaurs, or even a medium-sized tadpole.

Of course, all these sediments, with their time capsule contents of fossilized creatures from the past, were laid down long after the

Earth was formed and long after the decisive event took place in the chain of evolution—the origin of life itself in ancient seas. It is the rock from which those later sediments were derived—the primary bedrock of the Earth's crust—in which we are chiefly interested if we wish to date the Earth.

The key question remains: How old is the Earth? And to examine the answer that has come to be accepted on this score, we must look more closely at radioactive methods of dating.

CHAPTER 4

The Key to the Past?

I N THE YEARS FOLLOWING THE SECOND WORLD WAR, American chemist Willard Libby made a discovery that won him the Nobel prize for chemistry, which revolutionized the study of the Earth's prehistory, but which ultimately was to provide unexpectedly disconcerting evidence on the age of the Earth itself.

Libby's discovery was the now-famous radiocarbon method of determining the age of organic remains, which gave archeologists their first practical tool for routinely dating the past. At the time of its discovery and its first application to archeological sites around the world in 1949, the radiocarbon method appeared to confirm that humankind's past was indeed of great antiquity and that geologists and evolutionists had been perfectly justified in continually pushing further back in time the dawn of humanity.

Field archeologists in the 1950s, applying the new power given them by chemistry, confidently assigned absolute dates to early human prehistoric settlements with a precision that must have astounded their teachers of a generation before. The city of Jericho was said to have been a thriving human settlement 11,000 years ago, while Neolithic sites in Russia and Africa were dated as being well over 50,000 years old. The author of *Encyclopaedia Britannica*'s article on prehistoric Africa, for instance, says "Radiocarbon dating suggests that the Earlier Stone Age may have lingered on until about 55,000 B.C."

The readiness of science today to accept a great antiquity for the Earth and humankind contrasts sharply with the attitude of scientists little more than a century ago. This radical change in outlook involved the overthrow of the old geological belief in a catastrophic origin for the rocks of the Earth's crust and its replacement by the modern uniformitarian theory—the idea that the rocks have formed slowly over millions or billions of years.

At the time that Darwin set sail for South America in the *Beagle* in 1831, the Earth's age was reckoned merely in thousands of years, and not many thousands at that. One well-known early attempt to date the Earth is that of Archbishop James Ussher of Armagh, a noted Bible scholar who deduced through careful analysis of biblical texts that the Earth was created in 4004 B.C.. The Archbishop's finding was published in 1650 and soon after was added as a marginal notation to the Book of Genesis in the Authorized Version of the Bible where it remained until Victorian times, and can still be found occasionally today.

A contemporary of the Archbishop, Dr. John Lightfoot, Master of St Catherine's College and Vice Chancellor of Cambridge University, was able to endorse this date and indeed refine it with astounding precision. "Man was created by the Trinity," wrote Dr. Lightfoot, "on October 23rd 4004 B.C. at nine o'clock in the morning." As Ronald Millar has pointed out, only a Cambridge Vice Chancellor would have the audacity to assign the date and time of the creation to the beginning of the academic year.[1]

A number of the influential geologists in Darwin's day were also clergymen whose religious views strongly influenced their scientific beliefs. This religious complexion to geology in the eighteenth and early nineteenth centuries—an otherwise flourishing era for rationalist thinking in science—influenced theories of rock formation and the age of the Earth in two important ways.

First, widespread acceptance of the biblical creation story contained in Genesis meant that the cleric-geologists neglected to question how the Earth began or how life originated because they believed they already had the answers to these questions. And second, the creation story constituted a ready-made theory to accommodate all their scientific observations (often meticulously detailed) thereby stifling the formation of any new theory when they discovered new evidence in the field.

When these researchers found thousands of feet of compacted mudlike sediments containing the bones of dead animals, the discovery was taken as clear evidence of Noah's flood described in the Bible. Hence the prevailing geological theory of the pre-Darwinian era was that of catastrophism—the doctrine that the rocks of the Earth's crust were formed more or less simultaneously as a result of a divinely ordained Great Flood.

Some of the attempts by pre-nineteenth century geologists to fit their observations to biblical teaching appear obviously contrived and rather absurd from our perspective today. Swiss naturalist Johann Scheuchzer, who discovered some early vertebrate remains of a salamander around 1720, exhibited them widely as the remains of the imaginatively named *Homo diluvii testis*—Man, a witness to the flood. (Some believe that Scheuchzer was seeking to turn an honest copper or two with his discovery, in which case we must blame the gullibility of his customers rather than the inadequacy of eighteenth century science.)

In general, though, the observations of nature made at this time were models of scientific accuracy and would do credit to any modern researcher. Unfortunately when the theory of catastrophism fell into disrepute after Darwin, many of the observations of the cleric-geologists were rejected as religiously inspired prescientific thinking: observations that did indeed support a catastrophic origin for many rocks. The detailed evidence for catastrophism is examined in a later chapter, but one observation of this type that was well known in Darwin's day may be mentioned now by way of example: the occurrence of "graveyards" of millions of land-dwelling (not marine) creatures who suffered death simultaneously.[2]

Darwin and his supporters realized at an early stage that their theory demanded vast reaches of geological time to support the supposed microscopic changes in form from one generation to another. Equally, evolutionists stood in need of a geological basis for this great antiquity—a mechanism that worked slowly and gradually rather than one that worked suddenly and all at once. They rejected catastrophism and instead found the mechanism they sought in an idea taking shape among the new generation of secular geologists who asserted that sedimentary rocks (that is, fossil-bearing rocks) were formed slowly by the same processes that can be seen on the ocean bottom today: the deposition of silt and sand that becomes cemented and compacted over millions of years to form successive strata of rock.

Under the reassuring-sounding label of uniformitarianism these ideas were actively promoted by secular geologists like James Hutton and later Charles Lyell, who was Darwin's coach on geological issues. The uniformitarian doctrine is summed up in the famous phrase "the present is the key to the past"—a concept eagerly accepted by Darwinists as ready-made for their theory and one expounded on at length in Lyell's *Principles of Geology*, the primary geological work of the century, published between 1824 and 1833.

The important point to note here is that it was the imperative need for great antiquity that deposed catastrophism, rather than any new scientific discoveries or observations; it was a new way of looking at things, not a new piece of knowledge. But, superficially, the change in view seemed to be a shift away from naive belief in biblical tales of creation and flood, and toward a newly established scientific viewpoint. And those who continued to argue the case for a catastrophic origin of rocks were seen as merely making a last-ditch attempt to rescue the religious doctrine of the creation as told in Genesis.

Darwinists needed time, and lots of it: uniformitarians had the geological theory that demonstrated great antiquity. Geologists needed a firm foundation for the relative dating and correlation of the many sediments piled one on another in the past—the many strata of the geological column: Darwinists were able to supply the key to the stratigraphical succession of the rocks by comparative anatomy of the fossils contained in those strata, interpreted along evolutionist lines. Thus an unusual academic interdependence sprang up between the two sciences that continues to this day. A geologist wishing to date a rock stratum would ask an evolutionist's opinion on the fossils it contained. An evolutionist having difficulty dating a fossil species would turn to the geologist for help. Fossils were used to date rocks: rocks were used to date fossils.

A modern example of paleontologists using fossils to date rocks in a circular way is provided by one of the most famous of all North American dinosaur discovery sites: the rocks at Como Bluffs, Wyoming. I only regret that this example involves one of today's most innovative researchers, Robert Bakker of the University of Colorado. It was at Como Bluffs in the 1870s and 1880s that paleontologists such as Edward Cope and O. C. Marsh discovered more than 120 new species of dinosaur, including diplodocus and stegasaurus. The many strata exposed in the steep cliff at this seminal site have subsequently yielded many more specimens and they are still worked today by scientists from many universities.

Of the site, Robert Bakker says:

At a place like Como Bluffs you have layer after layer—it's like getting a burst of frames from a motion picture of how the dinosaurs came, flourished and went extinct. At any one place in the world, you don't have the whole history of dinosaurs, in fact you don't have the whole history of one family of dinosaurs, you just have a little burst of fossils.

We don't yet have radioactive beds that can give us a nice hard number [on the age of the deposit]. But by comparing the fossils we get at the bottom of the section and at the top, it's about 10 million years. So all of this history is played out roughly over about 2 million dinosaur generations, 10 million chronological years.

Ironically, not only is there no radioactive basis for the dating of Como Bluffs, there is, as Robert Bakker says, not even a complete history of a single dinosaur family at the site. Yet we are given the confident assertion concerning the number of dinosaur generations and the number of years to which this sequence is equivalent, with no solid physical basis. No other scientific discipline would be permitted even to consider such procedures, but when paleontologists date rocks by means of fossils, they do so with the authority of Charles Darwin himself.

This circular process ought to have aroused suspicion, if not among its practitioners then among scientists of related disciplines. In fact it went unremarked and unchallenged because the discovery and introduction of methods of dating based on radioactive decay in the early years of this century appeared amply to vindicate the Darwinist-uniformitarian view and to justify their interdependence.

In the last two decades, however, further research into these technical methods of dating has revealed a number of worrying inconsistencies in the now orthodox view of the Earth's age: radio-active dating techniques are far less reliable than was previously thought; the Earth could be much younger than has been supposed by Darwinists; and nothing like the billions of years required by evolution theory have elapsed since the Earth's formation.

The first clue that something may be amiss with the view of uniformitarian geology and its claim for an old Earth came para-doxically from the technique that seemed most to support that view—Willard Libby's radiocarbon dating method. To appreciate exactly why the radiocarbon technique has had such unexpected consequences, it is necessary first to look at just how the technique was supposed to work.

Radiocarbon—radioactive carbon 14—is a form of carbon cre-ated in the upper atmosphere by the bombardment of cosmic par-ticles from space. As radioactive carbon dioxide it permeates the atmosphere and passes into the bodies of plants and animals through the food chain. To any plant or animal, carbon 14 is indistinguish-able from the common carbon (carbon 12) which occurs naturally on Earth. Radiocarbon is relatively rare, so of the total amount of carbon in the body of a plant or animal only a minute fraction is radiocarbon. What makes this tiny fraction useful for dating, ar-gued Libby, is that the *proportion* of radiocarbon is the same for all living animals and plants the world over, and something that can readily be measured.

Radiocarbon begins to decay as soon as it is formed. When a quantity of radiocarbon is produced in the atmosphere, half of that amount will have decayed away (becoming nitrogen gas) in some 5,700 years. Half the remainder will decay in a further 5,700 years, and so on, until an immeasurably small residue remains. Once a plant or animal dies, it ceases to take in radiocarbon from the "terrestrial reservoir" or outside world, so the amount of ra-diocarbon in its body begins to dwindle through decay while the ordinary carbon remains unchanged. So, 5,700 years after a tree dies, it contains only half the proportion of radiocarbon to com-mon carbon that exists in a living tree, and in the living world in general. After a total of 11,400 years, or two half-lives, it will contain only one quarter the proportion in the outside world, and so on. After about five half-lives, or roughly 30,000 years, only an immeasurably small residue remains and so the radiocarbon test is only good for dating remains younger than this natural "ceiling."

To date an organic find (the test only works, of course, on the remains of once-living things, such as bones in a Neolithic burial, or Roman fence posts) it is only necessary to measure the amount of remnant radioactive carbon with a suitable counter and hence deduce when the specimen ceased to take in radiocarbon—when it died.

The great value of the test is that only a tiny fragment of an irreplaceable papyrus or rare skull is needed because it is the proportion of radiocarbon to ordinary carbon that is measured and compared with the proportions that exist in the terrestrial reservoir or living world today. In the end the whole technique rests, therefore, on knowing with some precision the ratio of radiocarbon to common carbon in the terrestrial reservoir today, and it was for making these measurements as well as developing the dating technique that Libby was awarded the Nobel prize.

There is just one further factor of some importance for the test to work properly: the standard mix of radiocarbon to ordinary carbon in the terrestrial reservoir must always have been the same throughout the lifetime of the test subject and in the years since its death. Take the case of archeologists setting out to determine the age of a Neolithic woman whose burial chamber they discover. If there had been a lot more carbon 14 around during the life of this early woman, the reading from her bones will be falsely inflated— she will appear a much more recent burial than she really was. Had there been a lot less radiocarbon around during her life, then the reading will appear falsely diminished and she will appear much older.

At the time that Libby and his co-workers were developing the new technique, in the 1940s, they had every reason to believe that the amount of carbon 14 in the world could not possibly have varied during the time that humankind had been on Earth simply because the Earth is of immense age, some 4,600 million years old. This great age stamps the radiocarbon technique with the seal of respectability because of what Libby called the "equilibrium value" for the radiocarbon reservoir.

After the Earth was formed and acquired an atmosphere, there would be a 30,000 year transition period during which carbon 14 would be building up. At the end of that period, the amount of carbon 14 created by cosmic radiation will be balanced by the amount of carbon 14 decaying away to almost zero. To use Libby's terminology, at the end of 30,000 years, the terrestrial radiocarbon reservoir will have reached a steady state.

Since the Earth, according to uniformitarian geology, is many, many times older than the 30,000 years needed to fill up the reservoir, then radiocarbon must unquestionably have attained equilibrium billions of years ago, and must have been constantly so throughout the few million years allotted to human history. To test this essential part of the theory, Libby made measurements of both the rate of formation and the rate of decay of radiocarbon. He found a considerable discrepancy in his measurements indicating that, apparently, radiocarbon was being created in the atmosphere somewhere around 25 percent faster than it was becoming extinct. Since this result was inexplicable by any conventional scientific means, Libby put the discrepancy down to experimental error.[3]

During the 1960s, Libby's experiments were repeated by chemists who had been able to refine their techniques after a decade or so of experience. The experiments demand almost heroic measures since the amounts of radiation involved are very small (only a few atomic disintegrations per second) and because of the need to screen out all other sources of radiation that would contaminate the result. The new experiments, though, revealed that the discrepancy observed by Libby was not merely experimental error—it did exist. It was found by Richard Lingenfelter that "There is strong indication, despite the large errors, that the present natural production rate exceeds the natural decay rate by as much as 25 percent . . . It appears that equilibrium in the production and decay of carbon 14 may not be maintained in detail."[4]

Other researchers have confirmed this finding, including Hans Suess of the University of Southern California, writing in the *Journal of Geophysical Research*[5] and V. R. Switzer writing in *Science*.[6]

Melvin Cook, Professor of Metallurgy at Utah University, has reviewed the data of Suess and Lingenfelter and has reached the conclusion that the present rate of formation of carbon 14 is 18.4 atoms per gram per minute and the rate of decay 13.3 atoms per gram per minute, a ratio indicating that formation exceeds decay by some 38 percent.[7]

The meaning of this discovery is described as follows by Cook: "This result has two alternate implications: either the atmosphere is for one reason or another in a transient build up stage as regards Carbon 14 . . . or else something is wrong in one or another of the basic postulates of the radiocarbon dating method."

Cook has gone one step further by taking the latest measured figures on radiocarbon formation and decay and calculating from them back to the point at which there would have been zero radiocarbon. In doing so, he is in effect using the radiocarbon technique to date the Earth's own atmosphere. And the resulting calculation shows that, using Libby's own data, the age of the atmosphere is around 10,000 years![8]

To anyone who, like me, was brought up on a diet of uniformitarian geology and Darwinian theory and to any high-school pupil or college student who opens a standard geology textbook, the suggestion that life on Earth may have a history as short as 10,000 years inevitably appears preposterous. Surely, the radiocarbon method has been tested against artifacts of known age and has been thoroughly vindicated? Surely the technique has been widely adopted in archeology with excellent results? And surely any fundamental flaw in the methods would have been discovered years ago?

It is perfectly true that radiocarbon dating has been tried on objects whose age is independently known from archeological sources and scored some impressive early successes. One of the very first artifacts to be tested was a wooden boat from an Egyptian pharaonic tomb whose age was independently known to be 3,750 years before the present. Radiocarbon assay produced the date of between 3,441 and 3,801 years, a minimum error of only 51 years. But after this promising start, the method quickly ran into difficulties. Anomalous dates were produced from later assays that showed that some living things may interact with parts of the reservoir that have been anomalously depleted of carbon 14 and thus appear to be much older than they really are.

In one of the most recent cases of anomalous dating, rock paintings found in the South African bush in 1991 were analyzed by Oxford University's radiocarbon accelerator unit which dated them as being around 1,200 years old. This finding was significant because it meant the paintings would have been the first bushman painting found in open country. However, publicity of the find attracted the attention of Joan Ahrens, a Capetown resident, who recognized the paintings as being produced by her in art classes and later stolen from her garden by vandals. The significance of incidents such as this is that mistakes can only be discovered in those rare cases where chance grants us some external method of checking the dating technique. Where no such external verification exists, we have simply to accept the verdict of carbon dating.

The position resulting from these anomalous discoveries was summarized by Hole and Heizer in their *Introduction to Prehistoric Archaeology*:

> For a number of years it was thought that the possible errors . . . were of relatively minor consequence, but more recent intensive research into radiocarbon dates, compared with calendar dates, shows that the natural concentration of Carbon 14 in the atmosphere has varied sufficiently to affect dates significantly for certain periods. Because scientists have not been able to predict the amount of variation theoretically, it has been necessary to find a parallel dating method of absolute accuracy to assess the correlation between Carbon 14 dates and the calendar.

The parallel dating method turned to in order to assess radiocarbon dating involves that strange tree the bristlecone pine, which grows at high altitudes in the mountains of California and Nevada and is the oldest living thing on Earth—some specimens said to be 5,000 years old.

The bristlecone pine has been exploited by Charles Ferguson of Arizona University to develop the science of dendrochronology—dating by tree rings. The tree is useful here because it lives to a great age and certain "signature" sequences of tree rings are said to be characteristic of specific years before the present, enabling a younger tree to be correlated with older trees (including dead ones) to stretch the tree-ring chronology further and further back. Cross-dating from one core sample to another by means of such signatures enabled Ferguson to construct a master chronology that spans a total of 8,200 years before the present. This has been used to check up on radiocarbon dating variations.

Hans Suess of the University of California in San Diego has radiocarbon dated the bristlecone pine samples of the master chronology and from this a table of deviation has been drawn up which in theory allows the inaccuracies of the radiocarbon method to be corrected for up to around 10,000 years ago.

Radiocarbon dating's inventor Willard Libby did not at first think that large deviations were possible. "When we developed the radiocarbon dating method," he said, "we had no choice than to assume that the cosmic rays had remained constant, though obviously we hadn't the slightest evidence that this was so. But now we know what the variations were."

Hans Suess was able to show precisely how variations in the amount of cosmic radiation changed the amount of radiocarbon in the atmosphere and his table indicates that by about 5,000 B.C., radiocarbon-derived dates are around 1,000 years too young.

"Whatever the source of radiocarbon," says Libby, "it mixes very rapidly with life on earth so we have a firm belief that the calibrations with the bristlecone pine apply worldwide."

Are archeologists happy with this result? In fact they appear rather confused by it. Before the bristlecone pine amendments, the dates given by radiocarbon dating had confirmed the widely held belief of diffusionists—that culture had spread from Egypt and the Middle East via Mycenae and Crete westward into Europe and then Britain. However, the new chronology indicates that, for instance, the island of Malta was carving spiral decorations and erecting megalithic structures *before* the supposed cradle civilizations further east. Many archeologists are unhappy about this, but the chronology now has the authority of both Libby and the dendrochronological corrections of Suess's bristlecone pine deviation tables.

A further difficulty has more recently been introduced into the controversy because the fundamental principle on which dendrochronology is based—that a tree ring forms each year—has been questioned. R. W. Fairbridge, writing on dendrochronology in *Encyclopaedia Britannica*'s entry on the Holocene epoch says:

> As with Palynology, certain pitfalls have been discovered in tree-ring analysis. Sometimes, as in a very severe season, a growth ring may not form. In certain latitudes, the tree ring's growth correlates with moisture, but in others it may be correlated with temperature. From the climatic viewpoint these two parameters are often inversely related in different regions.[9]

It is also possible for two tree rings to grow in a single year, when growth begins in spring but is later arrested by a period of unseasonal frosts and later starts up again.

These climatic variations presumably mean that a fresh set of correction tables will be needed to modify the bristlecone pine dates, although no one has yet devised a method of calibration for such tables. But whatever the outcome of the debate between archeologists and radiocarbon chemists, the key question for chemistry is how to explain the observed discrepancy between the rate of production of carbon 14 and its rate of decay in the atmosphere. Cook has suggested that one possible explanation of the discrepancy is that the atmosphere is still in nonequilibrium because the required 30,000 years have not yet elapsed since it was first formed.

Adherents of the old-earth theory have responded first by seeking to minimize the discrepancy—claiming that it is "around 10 percent" when it is really as great as 38 percent—and second by saying that the proportion of radiocarbon in the terrestrial reservoir may fluctuate over time and that we are currently going through a build-up phase. There is no scientific evidence to support this view but to someone who already believes in an old earth, the conclusion seems self-evidently more reasonable.

But what reasonable alternative could there be? How could the Earth possibly be merely thousands of years old? How could science have gone so far wrong?

CHAPTER 5

Rock of Ages

O NE DAY, MORE THAN TWENTY YEARS AGO, I picked up an apparently dull geology textbook and found my attention arrested by a single sentence. The book was called *Prehistory and Earth Models* and was by the professor of metallurgy at Utah University, Dr. Melvin Cook.[1] Cook, a physical chemist now in his eighties, is a world expert on high explosives and his textbook on explosives for mining is still a classic work of reference. Professors of metallurgy do not usually stir up trouble in the academic world, but what I had read in his geology book was more explosive than any text on TNT.

In his preface Cook wrote: "An attempt to publish a manuscript giving direct evidence for the short-time chronometry of the atmosphere and oceans entitled 'Anomalous Chronometry in the Atmosphere and Hydrosphere,' not unexpectedly nor without some cause, met with considerable opposition and was not published."

Who on earth had prevented Dr. Cook from publishing his paper? I wondered. And what could a metallurgy professor have to say that was so heretical that someone wanted to prevent its publication? I found that his book contained scientific evidence and reasoned argument which showed that something was terribly wrong with the orthodox scientific view of methods of dating. The most widely used methods, such as uranium-lead and potassium-argon, had been found to be seriously flawed, not merely in practice but in principle. In addition, the methods yielded dates so discordant as to make them unreliable.

Cook showed for example that if you used the uranium-decay method on the rocks of the crust you got the conventionally accepted age of over four thousand million years. But if you used the selfsame method on the atmosphere, you got an age of only a few hundred thousand years. He also showed that the entire amount of "radiogenic" lead in the world's two largest uranium deposits could be entirely modern. Clearly something was wrong.

When I dug deeper, I found that Cook was not a lone voice. Other papers by scientists in reputable scientific journals expressed similar doubts and findings. Funkhouser and Naughton at the Hawaiian Institute of Geophysics used the potassium-argon method to date volcanic rocks from Mount Kilauea and got ages of up to 3 thousand million years—when the rocks are known to have been formed in a modern eruption in 1801. McDougall at the Australian National University found ages of up to 465,000 years for lava in New Zealand that is independently known to be less than 1,000 years old.

I eventually came to the alarming realization that although radioactive decay is the most stable source of chronometry we have today, it is badly compromised as a historical timekeeper, because it is not the rate of decay that is being measured but the amount of decay products left. For this reason, all radioactive methods of geochronometry are deeply flawed and cannot be relied on with any real confidence in this application.

At the end of the last chapter, I asked, How could science have gone so far wrong? The answer turns out to be that it is not science which has gone wrong, merely those scientists seeking to defend a single idea—Darwinian evolution. Science has proposed many methods of geochronometry—measuring the Earth's age— all of which are subject to some uncertainties, for reasons I shall describe in a moment. But of these many methods, only one technique—that of the radioactive decay of uranium and similar elements—yields an age for the Earth of billions of years. And it is this one method that has been enthusiastically promoted by Darwinists and uniformitarian geologists, while all other methods have been neglected.

So successful has this promotional campaign been that today almost everyone, including scientists working in other fields, has been led to believe that radioactive dating is the only method of geochronometry worth considering, and that it is well-nigh unassailable because of the universal constancy of radioactive decay. In fact, none of these widely held beliefs is supported by the evidence.

To appreciate how and why radiometric methods are flawed, first look a little more closely at the problems which confront the geologist attempting to measure the Earth's age.

All methods of measuring time, whether for domestic or scientific purposes, rely on the same basic principle: monitoring the rate of some constant natural process. Today our most sophisticated chronometric methods involve the rate at which a quartz crystal vibrates when an electric potential is applied to it, and the rate at which radioactive elements decay—said to be the most constant source of all.

But having some readily available process to measure is not enough by itself. To measure elapsed time accurately we must be sure that the process does in fact remain constant, even when we are not watching. You must know the starting value of the clock— how much water was in your water clock to begin with or how tall your candle was before it was lit. And you must be sure that some external factor cannot interfere with the process while it is in operation, for instance, that a temporary power cut does not stop your electric clock while you are out walking your dog.

All these conditions apply to measuring time today. When it comes to the science of geochronometry, the process we choose will have started in prehistoric times, which we have no method of directly observing and verifying. This means we must make sure as far as possible that our three conditions were met in the past as well as in the present—and it is here that our problems begin.

Suppose, for instance, we were to take the increasing salinity of the oceans as a means of finding out how old the Earth is (a method actually proposed in 1898 by Irish geologist John Joly). On the face of it this is a promising method, since it can be assumed that initially the oceans consisted of fresh water, and the present-day accumulation of salt is due to erosion of land masses by rainfall and the subsequent transport of dissolved salt into the seas by way of the world's rivers. Even more encouraging is the fact that the rate of erosion of the land by rainfall is surprisingly constant each year— about 540 million tons of salt a year. All that would be necessary is to measure the present-day concentration of salt in the sea (32 grams per litre); calculate from this the total amount in all the oceans (about 5×10^{16} tons); and divide this total by the annual amount of salt deposited to get the age of the Earth in years.[2]

Using this method, Joly came up with an age of 100 million years. Unfortunately, when we apply the three conditions mentioned earlier to this method its shortcomings quickly become obvious. First, we cannot be sure that the annual runoff of dissolved salt has always been constant. Indeed there is good reason to suppose that climatic conditions have been very different in the past— with ice ages and major droughts for instance—and these conditions might have had an effect that is incalculable.

Second, we cannot be quite sure that there was zero salt in the sea to begin with. Initially, some salt might have been present, though no one can say how much, if any. (Recent research in the Atlantic suggests that salts may have been extruded into ocean basins from the molten magma beneath the crust.) And third, it turns out that an apparently constant process is interfered with by external factors. Large amounts of salt are recirculated into the atmosphere, and recent evidence suggests that the salt in the sea might actually be in a steady state—as fast as salt is deposited in the sea, it is picked up in the air and redeposited on land again. A large quantity of salt is evaporated by biological processes and still more is incorporated into bottom sediments through chemical processes, spoiling our "clock."

All methods of measuring the age of the Earth are subject, to some extent, to the same defects—quite simply, no one was there at the time to check up on our three criteria. The technique used by uniformitarian geologists to arrive at the tremendous age of 4,600 million years for the Earth is usually referred to simply as the "uranium" or "uranium-lead" method. Sometimes it is popularly referred to merely as radioactive or radiometric dating. The technique in question covers a family of methods involving the radioactive decay of a number of different metallic elements with very long half-lives (they stay radioactive for very long periods). These elements include uranium and its sister element thorium, which both decay into helium and lead; rubidium, which decays into strontium; and potassium, which decays into argon and calcium.

The basic principle is this: over very long periods of time uranium spontaneously decays into lead and helium gas. The rate of decay is remarkably constant. The atoms of the uranium are unstable and periodically throw out an alpha particle, which is the nucleus of an atom of helium. It is impossible to tell in advance when any particular atom will break apart in this way since the process occurs at random. But in any substantial mass of the mineral there will be many billions of atoms, and with very large numbers of events the "law of large numbers" operates to produce a statistically predictable result.

The important part of the theory is that the kind of lead into which uranium eventually decays is chemically distinctive from common lead already present in the rocks, and is referred to as radiogenic lead, a daughter product of the decay process. Common lead is an isotope called lead 204, while the decay product of uranium 238 is lead 206. In order to date a rock deposit a sample is taken and the amount of radioactive uranium, together with the amount of radiogenic lead it contains, is accurately assayed in the laboratory. Since the rate of decay is known from modern measurements, it is possible to calculate directly how long the uranium has been decaying—how old the deposit is—by how much radiogenic lead it has turned into.

The half-life of uranium 238 (one of the principal isotopes used) has been calculated to be 4,500 million years. To take a simplistic example, if the assay showed that a deposit was composed of half uranium 238 and half its daughter product lead 206, then one would draw the conclusion that the deposit was 4,500 million years old. (This, incidentally, is the average figure that is found for the Earth's crust although the figure is arrived at by extrapolation rather than direct measurement.)

On the face of it, uranium decay seems an ideal method of geochronometry, and above scientific suspicion. But, as in the case of radiocarbon dating, research in recent decades has begun to cast serious doubts on its reliability.

The first criterion for any method of geochronometry is that we must know the starting value of the process we are measuring; we must have a point of departure, or reference point, from which to make our calculations. On the face of it, uranium decay fulfills this requirement since the type of lead which results is said to be uniquely formed as a by-product of this process. If radiogenic lead—lead 206 and lead 207 from uranium, and lead 208 from thorium—really is uniquely formed as the end product of disintegration, then it is perfectly reasonable to suppose, as adherents of radioactive dating do, that there was zero radiogenic lead in the rocks of the Earth's crust when they first formed, and so we have a reliable starting point for our calculations. The same argument can be used to make us reasonably certain that no radiogenic lead could have intruded into the rocks by some other means, thus distorting the effects of the decay process.

But things are by no means as simple as they seem when investigated a little more closely. Cook has suggested there is another, and quite separate, mechanism by which common lead can be transmuted into a form which, on assay, will be indistinguishable from "radiogenic" lead. This transmutation can occur through the capture of free neutrons—atomic particles with enough energy to transmute common lead into so-called radiogenic lead. Where, though, could such a source of free neutrons be found? The answer is in a radioactive ore deposit such as uranium, where they occur through spontaneous fission!

In other words, the very process being measured can be moonlighting at another job. As well as spontaneously decaying into radiogenic lead, it is also making available a supply of particles which are simultaneously converting common lead into another isotope which, on being assayed, will be indistinguishable from a radiogenic product of alpha decay. Significantly, this is a mechanism that would tip our measurements in favor of an "old" Earth. Too much "radiogenic" lead would lead us to imagine that the process has been going on for much longer than it actually has.

In the neutron capture process, the isotopic values of lead would be systematically changed: lead 206 would be converted into lead 207, and lead 207 into lead 208. Interestingly, lead 208 usually constitutes more than half the lead present in any given deposit. This is normally interpreted as meaning that thorium, the parent element of lead 208, was very common in the deposit in question, although it could also be interpreted as indicating that free neutron capture is a far more important process in lead isotope formation than radioactive decay.

In *Prehistory and Earth Models*, Cook examined the lead content of two of the world's largest uranium ore deposits—in Zaire and Canada. He found that they contained practically no Thorium 232. However they do contain significant amounts of lead 208. This could have been derived only from lead 207 by neutron capture, says Cook, while all the so-called radiogenic lead can be accounted for on the same basis and the mineral deposits could be essentially of modern origin.[3]

Because Cook is a creationist as well as a scientist and because creationists have used Cook's findings as ammunition for their cause, strenuous attempts have been made by some scientists, such as G. Brent Dalrymple, a geologist with the U.S. Geological Survey, to discredit him and his research.[4] So far, however, neither Dalrymple nor any other dating advocate has offered a satisfactory explanation for the finding that there is practically no thorium 232 in the world's two largest uranium deposits, but that there are significant quantities of lead 208.

Dalrymple and others have asserted that the level of free neutrons available is too low to be capable of causing any significant change in the ratio of lead isotopes in deposits such as these. But if that assertion is correct then it becomes impossible to account on any rational basis for the quantities of lead 208 in Zaire and Canada.

So uranium decay fails the most important criterion for a reliable method of geochronometry. But it also fails a second criterion—that we must be reasonably sure no outside agency can interfere with the smooth running of our chosen process. Uranium does not naturally occur in metallic form but as uranium oxide. This material is highly soluble in water and is known to be moved away from its original deposit in large quantities by ground waters. The type of effect this has on dating is unpredictable since some parts of a mineral deposit can be unnaturally enriched while others are unnaturally depleted.

There is one further discovery relating to uranium dating that is of considerable relevance to attempts to measure the age of the Earth. As mentioned earlier, the final disintegration products of the decay process are two, not only lead but also helium gas. Like the lead which results from the decay process, the helium is also a radiogenic daughter product with an atomic weight of 4. In fact almost the entire amount of helium in the Earth's atmosphere is believed to be radiogenic helium, formed during the decay process throughout most of the Earth's history.

Now, if the uranium-lead dating technique were reliable, then the amount of this radiogenic helium in the atmosphere would yield a date for the Earth's age consonant with that yielded by measuring the amount of radiogenic lead in the crust. In fact, the dates are so different as to be irreconcilable.

If the Earth were 4,600 million years old, then there would be roughly 10,000 billion tons of radiogenic helium 4 in the atmosphere. Actually, there are only around 3.5 billion tons present— several thousand times less than there should be (0.035 percent to be precise).

Writing in *Nature* on the "mystery" of the Earth's missing radiogenic helium, Melvin Cook says,

> At the estimated 2×10^{20} gm uranium and 5×10^{20} gm thorium in the lithosphere, helium should be generated radiogenically at a rate of about 3 x 109 gm/yr. Moreover the (secondary) cosmic-ray source of helium has been estimated to be of comparable magnitude. Apparently nearly all the helium from sedimentary rocks and, according to Keevil and Hurley, about 0.8 of the radiogenic helium from igneous rocks, has been released into the atmosphere during geological times (currently taken to be about 5×10^9 yr). Hence more than 10^{20} gm of helium should have passed into the atmosphere since the "beginning." Because the atmosphere contains only 3.5×10^{15} gm helium 4, the common assumption is therefore that about 10^{20} gm of helium 4 must also have passed out through the exosphere, and that the present rate of loss through the atmosphere balances the rate of exudation from the lithosphere.[5]

Cook says that uniformitarian geologists have attempted to explain this discrepancy by assuming that the other 99.96 percent has escaped from the Earth's gravitational field into space—but this process has not been observed.

G. Brent Dalrymple has rebutted Cook's claim by suggesting a mechanism that might account for the missing helium 4. In his 1984 *Reply to "Scientific Creationism,"* Dalrymple says,

> Banks and Holzer (12) have shown that the polar wind can account for an escape of 2 to 4 x 10^6 ions/cm^2.sec of helium 4, which is nearly identical to the estimated production flux of (2.5 ± 1.5) x 10^6 atoms/cm^2.sec.[6]

There are two things that make Banks and Holzer's findings unsuitable for the purposes to which Dalrymple tries to fit them. The first is that the figure he cites for escape may be great enough to account for the production whose figures he gives, but that is only because he has selected a low estimate for production. In reality the escape rates he cites are not remotely great enough to account for the amount of helium 4 that must have been created and lost—remember we are looking for more than 10^{20} grams of missing helium. This means that if the Earth really is 4,500 million years old, then its atmosphere would have to lose helium at a rate somewhere around 10^{16} atoms/cm^2.sec., or some ten orders of magnitude faster than Dalrymple's figure, to account for the missing helium.

The second objection is that the figures he uses come from a time (nearly 30 years ago) when most space scientists assumed that the Earth was moving through the vacuum of space—that the atmosphere was surrounded by nothing but empty space. At that time it was believed that light hydrogen and helium atoms would either escape or be dislodged into the void.

More recent studies have suggested that far from losing helium, the atmosphere may actually be *gaining* quantities of this gas. As it orbits the Sun, the Earth moves not through empty space but through a thin solar atmosphere, which consists principally of hydrogen and helium resulting from nuclear processes within the Sun. Measurements in the upper atmosphere have suggested that the Earth is gaining helium by this means.

In his 1987 book *Gaia: A New Look at Life on Earth*, space scientist James Lovelock writes,

The outermost layer of the air, so thin as to contain only a few hundred atoms per cubic centimeter, the exosphere, can be thought of as merging into the equally thin outer atmosphere of the sun. It used to be assumed that the escape of hydrogen atoms from the exosphere gave the Earth its oxygen atmosphere. Not only do we now doubt that this process is on a sufficient scale to account for oxygen, but we rather suspect that the loss of hydrogen atoms is offset or even counterbalanced by the flux of hydrogen from the sun.[7]

Of course, Lovelock is writing about hydrogen not helium. However, helium is four times heavier than hydrogen and it is plentiful in the Sun's atmosphere since it is the principal product of the Sun's nuclear fusion process. If hydrogen is not lost but gained, then the same will be true for helium.

If we take the measured amount of helium 4 in the atmosphere and apply the radioactive dating technique to it, says Cook, we find that the calculation yields an age for the Earth of around 175,000 years. This procedure fails our criteria of reliability in that the possible acquisition of helium 4 from outside upsets the process.

The only conclusion that can be safely drawn from the discordance between the uranium-lead and uranium-helium dates is that this form of radioactive dating is unreliable.

What about the dating techniques based on other radioactive elements referred to earlier? The methods based on decay of potassium to argon and rubidium to strontium are also subject to some of the defects already described, as well as having specific problems of their own.

Potassium minerals are commonly found in many rocks. Potassium 40 decays by capturing an electron and turning into the gas argon 40, with a half-life of 1.3 billion years.

Advocates of the potassium-argon method claim that the argon gas that results from the decay of potassium 40 remains trapped in the crystal structures of the mineral in which it forms—"like a bird in a cage," to use Brent Dalrymple's phrase—and accumulates through the ages, thus acting as a clock when the stored daughter isotope is released and measured.

The potassium-argon method is suspect because the end product used for assay, argon 40, is a very common isotope in the atmosphere and the rocks of the Earth's crust. Indeed, argon is the twelfth most abundant chemical element on Earth and more than 99 percent of it is argon 40. There is no physical or chemical way to tell whether any given sample of argon 40 is the residue of radioactive decay or was present in the rocks when they formed. Moreover, as argon is an inert gas that will not react with any other element, its atoms will always be trapped in the crystal structures of minerals whether it is radiogenic in origin or not. Cook has calculated that even if the Earth were five billion years old, no more than 1 percent of the argon 40 currently present on Earth could be a radiogenic daughter product and it is thus highly probable that some of the argon 40 in all potassium minerals has been derived directly rather than as a result of decay.[8]

So, if radiogenic argon 40 is like "a bird in a cage," then it is a cage that already contains birds of the same feather, from which it is indistinguishable.

The possibility of anomalous inclusion of argon is not merely conjecture but is borne out by numerous studies of volcanic rocks that have resulted in false dates. Even modern volcanic lava formed in recent historical times has been dated as up to 3 billion years old by the potassium-argon method.

According to Noble and Naughton of the Hawaiian Institute of Geophysics:

> The radiogenic argon and helium contents of three basalts erupted into the deep ocean from an active volcano (Kilauea) have been measured. Ages calculated from these measurements increase with sample depth up to 22 million years for lavas deduced to be recent. Caution is urged in applying dates from deep-ocean basalts in studies on ocean-floor spreading.[9]

A similar study of Hawaiian basaltic lava actually dating from an eruption in 1801, near Hualalei, came up with potassium-argon dates ranging from 160 million years to 3 billion years.[10] In 1969, McDougall of the Australian National University measured the ages of lava in New Zealand and got an age of 465,000 years whereas the carbon dating of wood included in the lava showed it to be less than 1,000 years old. The suspected reason for the anomalous ages was the incorporation of environmental argon 40 at the time of the eruption, and the inheritance of argon 40 from the parent magma.

As well as the anomalous inclusion, or gain, of argon 40, it is also possible for mineral samples to become anomalously depleted of the gas if the rocks from which the sample comes have been heated after formation, for instance by further volcanic activity. Such disturbed samples will yield incorrect dates if a simple accumulation clock method is applied to them.

Dating advocates, such as Dalrymple, accept that potassium-argon methods can be flawed but claim that they know the occasions on which the results are correct and when they are incorrect: "Like all radiometric methods, the potassium-argon method does not work on all rocks and minerals under all geologic conditions. By many experiments over the past three decades, geologists have learned which rocks and minerals act as closed systems and under what geologic conditions they do so."[11]

The problem with this widely held belief is that there is no truly independent means of verifying the age of any given sample (other than the very exceptional cases mentioned above). And the experiments to which Dalrymple refers have consisted solely of rejecting dates that seem wrong while accepting those that seem right, "seem" in this context meaning in line with uniformitarian expectations, thus compiling a database of self-fulfilling predictions.

Radiogenic strontium—strontium 87—occurs in rocks as a result of decay of radioactive rubidium. However, this technique is again complicated by the fact that strontium 87 also occurs both as a daughter product of radioactive decay and as a commonly occurring element in its own right. Typically, rocks contain ten *times* more common strontium 87 than radiogenic strontium 87. Rubidium-strontium is also suspect because it is subject to exactly the same neutron capture process as uranium-lead. This time it is strontium 86 that can be transformed to strontium 87.

Most disconcerting of all is the fact that these various methods of dating commonly produce discordant ages for the same rock deposit. Where this occurs, a "harmonization" of discordant dates is carried out—in other words, the figures are adjusted until they seem right. The chief tool employed to harmonize discordant dates is the simple device of labeling unexpected ages as anomalous and, in the future, discarding those rock samples that will lead to the "anomalous" dates. This practice is the explanation of why many dating results seem to support each other—because all samples that give ages other than expected values are rejected as being "unsuitable" for dating.

If radioactive dating is seriously flawed as claimed here, why is it so enthusiastically embraced by dating scientists and so readily accepted by their academic colleagues?

On the face of it, radioactive dating is the most accurate source of chronometry available. Indeed, our most trustworthy timepieces are atomic clocks: clocks regulated by precisely the same processes used in dating techniques. And because radioactive decay is the most stable process known, then it appears that methods of geochronometry based on radioactive decay must themselves be the most accurate methods.

This widely held view fundamentally misrepresents the true nature of radioactive decay geochronometry. The accuracy of such techniques is not only critically dependent on the constancy of the rate of decay, but it is even more critically dependent on the accurate assay of the residue of the decay process—how much argon 40 is left or how much strontium 87 is left—and how that residue is distinguished from the nonradiogenic argon 40, or strontium 87, that occurs naturally in the same rocks.

This issue has nothing to do with how constant radioactive decay processes may be: it is purely a human problem in measurement. If the scientist conducting the experiment fails to measure the residue accurately, the age he gets will be distorted by an unknown number of years.

But how is it possible that dozens of scientists around the world involved in dating techniques could all be misled on such fundamental matters? How could so many scientists be wrong?

I believe there are at least four ways in which dating scientists could mislead themselves: ways that may be transparent to them, and which could lead them to obtain comparable results apparently independently.

First, there is the untestable error. When errors in radiometric dates are pointed out by critics, advocates of the method usually dismiss such criticism on the grounds that errors are very rare in comparison with the thousands of dates that are not found to be incorrect. This is a misleading argument because the overwhelming majority of dates could never be challenged or found to be flawed since there is no genuinely independent evidence that can contradict those dates. The reason why known anomalies are very rare is simply because independent evidence is very rare.

What is alarming is that in the very few cases of truly independent evidence we have—such as Mrs. Ahrens's rock paintings, and the volcanic lavas in Hawaii and New Zealand—the measured dates are spectacularly wrong. The response of radioactive dating advocates is to reject the few cases of independent verification as aberrations, and to prefer instead their theory purely because of its *internal* consistency, principally that it fits with a belief in an old Earth. In doing so, they are rejecting the only real independent check available.

Second, there is the phenomenon of "ballpark" thinking. This is exemplified by the error that was made in the curvature of the mirror for the Hubble space telescope. The error was not discovered by normal inspection processes, even in one of the world's best-equipped laboratories, because it was so big—more than a centimeter out—that it was outside the range that anyone was mentally prepared to check on. Had it been a millionth of a meter out, it would have been spotted at once.

Ever since Charles Lyell estimated that the end of the Cretaceous was 80 million years ago, the accepted value has been in this ballpark. Any dating scientist who suggested looking outside the ballpark, at 20 million years or 10 million or 5 million, would be looked on as a crackpot by his colleagues. More significantly, perhaps, he would not be able to get any funding for his research.

A third potential source of error is the phenomenon of "intellectual phase-locking." It is not widely realized that the published value of physical constants often varies. Before it was settled internationally by definition, the measured value for the velocity of light varied considerably, as did the gravitational constant and Planck's constant. One reason for such variation is that all scientists make experimental errors that they have to correct. They naturally prefer to correct them in the direction of the currently *accepted* value thus giving an unconscious trend to measured values. This group thinking has even been given a name: "intellectual phase-locking."

Fourth, there are powerful professional pressures on scientists to conform to a consensus. Dating geologists are offended by the suggestion that their beliefs can or would influence the dates obtained. Yet nothing could be easier or more natural. Take for example a rock sample from the late Cretaceous, a period which is universally believed to date from some 65 million years ago. Any dating scientist who obtained a date from the sample of, say, 10 million years or 150 million years, would not publish such a result because he or she will, quite sincerely, assume it was in error. On the other hand, any dating scientist who did obtain a date of 65 million years would hasten to publish it as widely as possible. Thus the published dating figures always conform to preconceived dates and *never* contradict those dates. If all the rejected dates were retrieved from the waste basket and added to the published dates, the combined results would show that the dates produced are the scatter that one would expect by chance alone.

Dating scientists have looked for a technique that would enable them to eliminate the problems of the simple accumulation clock method caused by inclusion or depletion of daughter isotopes. They believe that they have found such a technique in the idea first proposed by L.O. Nicolaysen of Witwatersrand University in 1961 and which is usually called the Isochron technique.

Geologists said to themselves, if we can find a way of using not just a single isotope, but of linking together several isotopes, and if we get a concordance of ages when we measure the linked group, then we can have a high level of confidence that the age we obtain is real and not a disturbed date. The main reason for believing this is that both of the two disturbing phenomena—inclusion or depletion of daughter isotope—will affect the different isotopes in a rock sample *differentially*, so they can no longer be made to lie on the same straight line when their ages are plotted on an Isochron graph.

On the face of it, the Isochron technique solves the basic problem of the simple accumulation clock method. In reality, it solves it only in a single limiting case—the case where all daughter isotopes are measured with perfect accuracy. If there is any *systematic* reason why the assay of the daughter isotopes is flawed, then the Isochron method is worse than useless—it is actively misleading, because it will cause geologists to place a high level of confidence in results that are actually false.

But, of course, the whole problem with radiometric methods is the difficulty of making accurate assays of the daughter isotopes coupled with the fact that there are a number of pressures compelling geologists to arrive at certain acceptable target dates and reject unacceptable dates in their published results.

In reality the apparent concordance of some of the dates derived by Isochron radiometric techniques is an artifact of two influences: the selection of "suitable" rock samples for assay and the rejection of "unsuitable" samples; and the selection of only some dates for publication and the nonpublication of others as being erroneous.

That the Isochron technique does not, in practice, provide the high level of confidence that some geologists attribute to it can be seen in the case history examined later in this chapter.

In evaluating the strength of the evidence and arguments against radiometric dating, the sticking point for many reasonable people is that a great age for the Earth—in the region of 4,500 million years—seems securely arrived at, whatever lesser problems may remain to be ironed out in radiometric dating techniques. Yet, as Melvin Cook has pointed out, the Earth may be made of materials that are 4,500 million years old and yet still have been formed relatively recently. Even if dates for meteorites and other celestial bodies such as the Moon and Mars could reliably tell us the age of the materials comprising the solar system, they still cannot tell us when the Earth itself was formed.

No part of this book has attracted such heated and vigorous rebuttal as this chapter on the flawed nature of radiometric dating. Advocates of radiometric dating have said that it is wrong of me to charge that discordant dates can be derived for the same deposit by different radiometric methods, wrong to say that such discordant dates are harmonized in the laboratory and wrong to say that dating scientists would be confused by the anomalous presence, or absence of, for example, nonradiogenic argon 40. One critic wrote to say that it is "dishonest" of me to include examples such as the modern Hawaiian lavas. "This is the sort of thing that is allowed for in radiometric dating," he told me indignantly. Another critic wrote and told me that the use of "Isochron" techniques for radiometric dating ensures that spurious dates would be eliminated and lead to a high level of confidence in radiometric dates.

These beliefs are no doubt sincerely held, but to show just how misguided they are, let me give a brief summary of one episode—involving some of the world's most distinguished isotope-dating laboratories—that embraced all the dating errors referred to above, despite every precautionary measure and attention to detail.

Paleontologists have made many important discoveries of human bones and tools at Lake Turkana (formerly Lake Rudolph) in Kenya. Among the deposits in which important finds have been made are those marked by a layer of volcanic ash or tuff identified by Kay Behrensmeyer of Harvard and which has become known as the KBS (Kay Behrensmeyer Site) Tuff.

From as long ago as 1967, when Richard Leakey began making finds there, it became important to try to date the KBS Tuff. Although it is volcanic and hence promising for the potassium-argon method, the deposit is not "juvenile" or original but has been transported by water and laid down as a sedimentary rock. It thus contains some foreign material including much older particles that could give an anomalous date—a fact which geologists who have dated it have recognized and which they have dealt with by selecting suitable juvenile particles to date.

In 1969, F. J. Fitch of Cambridge and J. A. Miller of Birkbeck College, London, dated the KBS Tuff as "very close to 2.6 million years old."[12] This had important implications later because when Richard Leakey found a very rare human skull below the KBS Tuff, he was able to say that it was found below rock that was "securely dated" at 2.6 million years ago.[13]

In 1976 *Nature* carried a second article by Fitch, Miller, and Hooker. They had refined their 1969 date using a more accurate constant of decay and found an age of 2.42 million years ago. In the same paper, the authors referred to "a small programme of conventional total fusion potassium-argon age determinations on East Rudolf pumice samples undertaken at Berkeley."[14]

The experiments they referred to were conducted by G. H. Curtis and colleagues at the University of California at Berkeley who, using potassium-argon dating, came up with dates of 1.6 and 1.82 million years for the KBS Tuff—a discrepancy with Fitch's results ranging from half a million years to close to a million years.[15]

Commenting on the discordant dating, Fitch said, "Potassium-argon apparent ages in the range 1.6–1.8 million years obtained from the KBS Tuff by other workers are regarded as discrepant, and may have been obtained from samples affected by argon loss."

What is especially interesting about these results is that both teams used Isochron methods—the methods that are claimed to *ensure* mistakes cannot be made simply because of anomalous loss or gain of argon, as in the Hawaiian lavas. Thus Fitch was alleging that the Berkeley team had got their sums wrong precisely because they failed to allow for argon loss—the very fault that my critic assured me was "the sort of thing that is allowed for in radiometric dating."

Perhaps because the issue of discordance had become public, Fitch went even further in his *Nature* paper and disclosed that the Berkeley group reported "scatter" in their dates ranging from 1.5 to 6.9 million years, a range large enough to cast some doubts on the accuracy of their work. By comparison, in their own experiments, Fitch and his colleagues claimed much lower "scatter" in apparent ages ranging from 0.5 to 2.4 million years implying that their measurements were more accurate.[16]

The controversy was brought to a close in 1981 by an argon-40-to-argon-39 study by Ian McDougall of the Australian National University, giving a date of 1.88 million years. As this was halfway between the two previous discordant studies, the combatants decided to call it a day—even though it meant they were both wrong by a large margin.[17]

In his paper McDougall frankly confessed that "conventional potassium-argon, argon-argon and fission track dating of pumice clasts within this tuff have yielded a distressingly large range of ages."[18]

Indeed, McDougall went even further than this rare emotive statement, because he revealed that the "scatter" referred to by Fitch was in reality even greater than that of Curtis. Fitch and Miller actually reported results of ages ranging from 0.52 to 2.64 million years for one set of samples and ages from 8.43 to 17.5 million years on another sample before eventually settling on their 2.6-million-year date.

McDougall concluded, "On the basis of the large scatter in the ages and the small proportion of argon-40 in the gas extracted from the anorthoclase concentrates, I suggest that the results are analytically less precise than given by these authors."

In the restrained diplomatic language of science, this is the equivalent of one scientist whacking another over the head with the sort of club that Lake Turkana Man was probably using on his enemies anywhere between 0.5 and 17.5 million years ago.

One aspect of this affair that deserves special attention is that all the scientists dating the formation started by selecting rocks they thought were the right age and discarding samples which seemed wrong. No one doubts that this is done honestly and intelligently. But the question must be asked: How do dating scientists know in advance which are the right rocks and which the wrong rocks? What *scientifically* led them to reject dates of 0.5 million years or 17.5 million years in favor of 2.6 million?

The answer that dating adherents give is that any scientist would exclude the few extreme measurements and settle for the majority of figures that are clustered together in a straight line or "plateau" when the results are plotted graphically. But, of course, had they measured the salinity of the oceans as a means of geochronometry (a method which as we saw earlier is known to be flawed) they would have found the same kind of "plateau" grouping for most results, regardless of where they took their seawater samples, because the method itself is systematically flawed. The majority of their dates would have been in the range of 100 million years because that is what the current salt content and annual erosion figures indicate. Constancy of results is not an indicator of correctness when the method itself is defective.

The truth is that, to those who dated the KBS Tuff, the chosen date of 2.6 million years seemed to be more "reasonable" than 0.5 million or 17.5 million. And the word reasonable in this context can be interpreted only as meaning consistent with uniformitarian and Darwinist beliefs on dating. The objection to this viewpoint is that "being reasonable" is not an acceptable substitute for scientific measurement and proof.

The fact is that presently it is impossible to say with any confidence how old the Earth is, beyond the obvious fact that it predates the calendar of human history.

PART TWO

CLAY

CHAPTER 6

Tales from
Before the Flood

In 1922 archeologist Leonard Woolley began to excavate the remains of one of the world's oldest cities, located between the Tigris and Euphrates rivers in Mesopotamia, or present-day Iraq. Woolley's hopes of great discoveries at the site of the biblical city of Ur were more than fulfilled. But what he found not only caught the archeological world by surprise, it also sent a ripple of consternation throughout the world's natural history and geological museums.

Six thousand years ago civilization arose in the plains of Sumeria, where many famous cities flourished and died. It was here that the legendary kings of Babylon lived and here that writing was invented. The fame of Ur has outlasted many other Sumerian cities because the Bible gives it as the birthplace of the patriarch Abraham—"Ur of the Chaldees."

The site of the city was identified at the end of the last century as present-day Tell al-Muqayyar ("the mound of pitch") through clay cylinders inscribed in the Akkadian language. Woolley's expedition was sent out by the British Museum after the First World War to examine and report on the remains.[1]

In common with other cities of the Sumerian plain, Ur today is little more than a gigantic mound of rubble: ruin piled upon ruin as each generation simply constructed new houses and public buildings directly on top of old ones which had crumbled with age, or which were knocked down to make way for newer property developments. By cutting trenches straight down through the mound, Woolley planned to reveal a slice of the history of Ur, its people and their artifacts. Eventually, his excavation took him so deep he found material from a period of immense antiquity which actually predated the Sumerian people and which he named the "al-'Ubaid period."

Driving deeper still, Woolley hit what most of his workers took to be the end of their dig, a thick bed of clay and silt. But continuing to dig, he passed through the thick bed of water—laid sediments and emerged again into the remains of civilised life, including Al-'Ubaid pottery.

He had clearly found the remains of a great flood that had inundated the Al-'Ubaid people, temporarily obliterating their community until it began to flourish once more. "No other agency could possibly account for it," wrote Woolley. "Inundations are of normal occurrence in lower Mesopotamia, but no ordinary rising of the rivers would leave behind it anything approaching the bulk of this clay bank: eight feet of sediment imply a very great depth of water, and the flood which deposited it must have been of a magnitude unparalleled in local history." Woolley believed he had found evidence of the Great Flood described in the Bible, and the evidence for this is persuasive.

The flood sediments he discovered date from around 3,000 B.C., early in the establishment of civilization in the area. Sumerian clay tablets from around 2,000 B.C. give an account of the flood as being divine retribution from a Sumerian god. The deity, however, takes pity on one man, Uta-Napishtim, and tells him to construct a boat, making it watertight with pitch. Uta-Napishtim saves his family and many animals aboard his boat, which survives seven days of rain. At the end of this week, Uta-Napishtim sends out a dove and a swallow, which return to the boat. Later he sends out a raven that does not return because it has found dry land. The boat comes to rest on a mountain top.

Some archeologists think that the Noachian flood story in the Hebrew Bible was borrowed by them from their neighbors the Sumerians. This idea is said to be supported by the fact that while there is ample evidence of such a flood in Sumeria, none has been found in the lands occupied by the people who wrote the Hebrew Bible.

The extent of the Sumerian flood was very substantial: a deposit 8-feet thick covering an area some 400 miles long by 100 miles wide—a total of many billions of tons of material. And it was this discovery that sent a buzz through the corridors of uniformitarian geology. For here, at last, was evidence of a real *Homo diluvii testis*—man a witness to the flood.

Because this catastrophic event had occurred within recorded history then—uniquely in the geological record—here was direct evidence of a substantial sediment that must have been laid down rapidly and all at once, rather than slowly over millions of years. And if this stratum, then why not others? If, as Hutton believed, the present is the key to the past, then was not this sediment the key to previous rock formations?

At almost the same time that work began on the excavation of the biblical Ur, a German meteorologist, Alfred Wegener, published a theory that was greeted with universal derision by the world scientific community—continental drift. Wegener's idea—that the major land masses were once joined, but have subsequently been forced apart—was regarded by geologists little more than forty years ago as belonging to the lunatic fringe of pseudoscientific beliefs. Yet, since the 1960s the evidence for continental drift has become overwhelmingly convincing and today few doubt its validity.[2] Perhaps to disguise their embarrassment at rejecting the idea so scornfully in the past, uniformitarian geologists have made continental drift respectable by rechristening it as "plate tectonics"—a subject now included in all geological curricula and textbooks.

In its present-day form, the idea is that the continents are the visible portions of gigantic "plates" whose edges are largely concealed beneath the oceans or deep in the Earth's crust, and which are "floating" on the semifluid material of the Earth's mantle. The continents are thus rather like pieces of cracked eggshell, floating on a soft-boiled egg.

The reason that continental drift was so disreputable to early twentieth-century geologists was that the forces required to crack the Earth's crust apart must have been cataclysmic, and this awakened the suspicion of Darwinists that catastrophism was not only about to rear its ugly, dinosaurlike head once more but was actually to gain admittance to the scientific drawing room through the back door.

The chief evidence for continental drift is the complementarity of geological features and coastal outlines of the continents; the apparent wandering of the magnetic poles along different paths for different continents; and the young ages of marine sediments and the ocean floors. When allowance is made for their continental shelves, the Atlantic shorelines of Africa and South America appear to be pieces of a former whole as do those of Europe and North America. Studies of paleomagnetism show that at some time in the remote past the rocks of the crust "pointed" to a different North Pole and have since moved (although some geologists remain skeptical on this issue). Drilling near ocean ridges in the Atlantic shows that the sediments overlying the bedrock get older as you move away from the ridge crest from where it is thought a continental plate is spreading.

It is now widely accepted that the land masses that have been pushed as much as a thousand miles apart probably did form a single land mass, which Wegener christened "Pangaea" (all-Earth). Today the major (northern) portion of this land mass is usually called Gondwanaland after rock strata in India which can also be found in South Africa and South America.

The important question for geology, of course, is just what agency caused the original continent to break apart? From the standpoint of evolution theory, an equally important question is, Precisely when did this event take place? While there has as yet been no agreement on the cause of the event, uniformitarians have predictably dated its occurrence as having happened during the Mesozoic era, which they believe lasted from 250 million years ago to 65 million years ago.

The central problem with continental drift, or plate tectonics, and the factor that delayed its acceptance for decades is that no one so far has proposed a satisfactory mechanism to drive the process. A number of explanations of the cause of continental drift have been proposed, each with its merits and difficulties. They include: tidal forces; expansion of the Earth; convection currents in the semifluid mantle; and successive loading and unloading of the crust (by glaciers, for instance).

A satisfactory model of the process has to meet three main criteria. First, it must demonstrate a mechanism that can produce sufficient force to initiate the breaking-apart of the crust. This requires very large amounts of energy to be released, whether the fracture is caused by compression (like squeezing an egg in your hand) or by tension (like holding a telephone directory by its edges and pulling it apart).

Second, the proposed mechanism must also provide sufficient energy to drive the fractured "plates" apart, in some cases riding over neighboring plates, and in other cases pushing directly against the edges of adjacent plates, folding them to form mountain ranges and, perhaps, thickening the crust in places. This is an important consideration because this "shouldering aside" of whole continents and mountain-building activity required even more energy than cracking the crust apart in the first place. Finally, the proposed mechanism must provide sufficient energy to break the crust and part the continents, but at the same time it must not generate excessive amounts of heat. Although it was once believed that the Earth was cooling as its molten interior lost heat, it is now known that the Earth's overall temperature is roughly constant, since heat loss from the surface is balanced by heat generated within the crust by radioactive decay. The continental drift mechanism must not disturb this heat balance.

Looked at from this point of view, most of the mechanisms proposed fail to account satisfactorily for the distribution of land masses that we observe today. Melvin Cook has made a detailed study of the most promising models and has shown that they are incapable of providing the required energies.

Probably the hypothesis most favored today by uniformitarian geologists is that continental drift is due to mantle convection currents. Underneath the crust, the Earth's mantle is subjected to intense heat and pressure and under these conditions behaves like a semifluid. A rough analogy is the molten iron poured from a blast furnace, with a crust of solid slag floating on top. Heat currents rise through the mantle from the core, travel for some distance along the base of the crust losing heat, and then descend, causing a vast circular movement of the mantle in the vertical plane.

It is hypothesized that the heated mantle material in one circular current may cause friction against the base of the crust, dragging it apart from an adjacent section of crust, which in turn is being dragged in a different direction by an adjacent mantle convection current.

Cook has calculated that the amount of heat generated by mantle convection great enough to cause continental drift would be between 1,000 and 10 billion times greater than the rate of radiogenic heat generation in the crust as a whole. "Clearly," says Cook, "such currents are impossible because either they would melt the Earth in a very short time, or one would observe an enormously greater heat flux from the Earth than is actually observed."[3]

There are other objections to convection currents, too. The measured rate of flow is far below the velocity required by theory for them to be capable of breaking and shifting the continents. In addition, the "velocity-gradient" of the viscosity of these currents (how runny the semifluid material is) would have to be at least 100 million times greater than currently postulated in order to cause continental drift.

Another theory enjoying some popularity in recent years is that the Earth may have expanded, thus cracking the original Pangaea apart. An advantage of this model is that it would also explain the observed expansion of the ocean basins. Although at first sight the idea of an expanding Earth seems rather farfetched, the theory does have considerable merit. An expansion in the surface area of the Earth of about 45 percent would account for the separation of Pangaea into today's fragments. To get a 45 percent increase in surface area would mean an increase in the Earth's diameter of about 20 percent.

Naturally, true to uniformitarian principles, this expansion is deemed to have taken place over immense reaches of time, during the 250 million years that are said to have elapsed since the Paleozoic era. This would mean an increase of around one centimeter a year in the Earth's diameter—on the face of it, not an unreasonable amount.

Unfortunately, the problem with this idea arises from exactly the same defect—its energy requirement. To fuel the expansion would take the entire chemical energy bonding together all the matter comprising the Earth. Chemical changes can thus be ruled out as being responsible for terrestrial expansion. The expanding Earth idea lacks a mechanism. In very recent times, however, a further attempt to rescue this idea has been made by postulating the steady-state creation of matter in the Earth's core as fueling the expansion. The idea that hydrogen atoms might naturally be continually coming into existence was first made popular by astronomer Fred Hoyle who suggested the process might be occurring in the space between the stars. The difficulty here is that the idea is confined wholly to the realm of theory since no one has ever observed or measured the steady-state creation of matter either in the Earth's core or in interstellar space.

Perhaps one of the main stumbling blocks to geologists in arriving at a satisfactory mechanism of the breakup of Pangaea and its redistribution around the Earth has been the persistence of the idea of continental "drift." Loyally obeying uniformitarian principles, geologists have looked for a mechanism that would act slowly and gradually to part the continental land masses. Indeed, when Wegener proposed the concept in 1912, he wrote of "wandering" continents, with its connotation of land masses floating gracefully about like icebergs on a calm ocean. But, as Cook has pointed out,

> One can be sure that continents cannot really simply wander aimlessly over the surface of the Earth: exceedingly strong forces must be applied to cause them to move through the powerful ocean crust. In fact when they do move it is only under a force sufficient to fracture and plastically deform massive rocks of extremely high strength, a process that cannot occur uniformly but only in certain sudden, explosion-like processes.[4]

In developing a non-uniformitarian model, Cook has taken up the suggestion of Hapgood and Campbell[5] that thickening of the polar ice caps places stresses of the required magnitude on the land masses beneath. In Cook's model, Pangaea stretched from pole to pole. Buildup of ice at one or both poles finally snapped the crust and the corresponding pressure at the other pole helped determine the direction of the main fracture. The pressures on the crust were thus rather like those on the shell of a hard-boiled egg being squeezed at both ends in a vice. The fracture would have occurred rapidly and explosively and the subsequent moving apart of the newly formed continents would also have been a rapid, rather than slow, drift.

There is evidence, says Cook, that the Wisconsin ice cap (the ancient Arctic) suddenly disappeared roughly 10,000 years ago. This mass of ice amounted to some 100 million billion tons. For it to have been melted by the Sun would have taken a minimum of 30,000 years even under ideal conditions. But this would have been more than enough for the Earth to "recover" naturally from being deformed since the "relaxation time" of the Earth's crust is less than 10,000 years. In order to account for the persistence of the Wisconsin depression it is necessary to conclude that the ice was dissipated catastrophically. According to Cook, independent evidence confirms that, in that region, the Earth's crust rapidly began to uplift at the same time as the ice disappeared, some 10,000 years ago. The mass of ice and snow thus released formed the present Arctic and Atlantic oceans, whose water content agrees reasonably well with the mass of the Wisconsin ice cap calculated from crustal depression data.

The ice-cap model does not have a direct bearing on the age of the Earth, since it can be argued that even if the continents did break up as recently as 10,000 years ago, the Earth might still be of very great antiquity—perhaps billions of years as Darwinists believe. But the model does have an important indirect bearing on methods of geochronometry because it challenges a key part of the Darwinian view of historical geology.

Darwinists believe that when the continents parted, more than 65 million years ago, the primitive mammals then in existence were separated geographically into distinct populations. In isolation these populations are said to have evolved quite independently to become, on one hand, the marsupial mammals of Australasia and, on the other, the placental mammals of Europe and America. This process is said to have led to some remarkable similarities. An often-cited example is the similarity of the Tasmanian marsupial wolf and the American and European timber wolf, which are non-marsupial.

This parallel development or "convergence" is seen by many Darwinists as important evidence in favor of the natural selection mechanism (and is an issue examined in detail in chapter 16). Clearly, however, if only some 10,000 years (or even 100,000 years, or 1 million years) have elapsed since the continents have parted, then nothing like enough time has passed for any appreciable evolutionary change to have taken place by means of spontaneous genetic mutation, and Darwinists can no longer appeal to the separation of land masses to support their theory. They have also to account for the similarity of Australian marsupials and their placental counterparts elsewhere by some other mechanism.

So it is reasonable to say that the ice-cap model points to a "recent" origin of life because it dramatically reduces the time scale in which certain key phases of evolution were formerly supposed to have occurred and rules out a mechanism that relies on random mutation.

As far as Darwinist theory is concerned, examples such as the Sumerian flood deposit, and the possible sudden dissolution of the Arctic ice cap, have a special significance. Darwinists reject any geological findings that point to catastrophic rather than gradual formation of rocks, for they threaten to reduce dramatically the historical time scale available for evolutionary processes. Yet such rejection is surprising in light of the geological evidence that contradicts the idea of slow, gradual formation.

CHAPTER 7

Fashioned from Clay

THE SYNTHETIC THEORY OF EVOLUTION RESTS not on seven pillars of wisdom, but on a solitary monolithic support—the geological column that is displayed in textbooks, classrooms, and natural history museums around the world.

As a teaching aid, as a powerful multilayered symbol of world prehistory, above all as a public relations tool for the general theory of evolution, the geological column has been a brilliant success. It has the answer to every question on evolution and the age of the Earth; it is the one thing every schoolchild takes home from museum field trips.

The geological column is both extremely simple and extremely complex. To begin with, the column was simply intended to represent the rocks of the Earth's crust, in sequence and roughly in scale with the age of rock, using colors to represent rock types in much the same way that the London underground or New York subway are represented by colored lines on the tube map.

Like the tube map, the geological column has taken on a meaning that is more literal than symbolic. The first change occurred when *relative* dates were added. The colored layers ceased to represent rock formations and became historical periods. Uniformitarians began to talk about the Cambrian "period" or the Cretaceous "period" instead of the Cambrian rocks or Cretaceous rocks. The next step was to assign *absolute* dates to various horizons within the column, for instance the primary rocks at the very bottom and the volcanic or igneous intrusions which appeared from time to time. And once the absolute dates were in place, and the sequence of rocks had become a chronology of the Earth, it was a natural final step to include family trees showing how the animal and plant kingdoms were related through common ancestors dominant in the various periods—complete with dinosaurs and ape-men.

The confusing relationship between the substantive role of the geological column and its role as an evolutionary metaphor is so subtle that it often escapes notice entirely. The first and most important of these confusions occurred when historical order was assigned to the many-colored rock strata. For this simple act carried profound implications about *how* and *how quickly* those rocks were formed. If the Cretaceous "period" lasted for 65 million years as evolutionists believe, then the chalk which was laid down during that period must have accumulated very slowly.

Some three quarters of the Earth's land mass is covered by successive layers of sedimentary rocks—that is, rocks like the chalk laid down under water and sometimes enclosing fossils. (The term "rock" is used by geologists to denote not only hard substances like limestone or sandstone, but also clays, shales, gravels, sands and any other substantial deposit of waterborne material.)

The conditions under which these deposits have been laid down are said to be analogous to the conditions which exist today; ranging from the ocean bottom, to the floor of shallow lakes; to coral reefs; to rivers and their estuaries or deltas. Conditions include both salt water and fresh water; tidal and nontidal; inland sea and open ocean.

Some sedimentary rocks are composed of pieces of the primary or volcanic rocks that originally formed the Earth's crust. These pieces range from boulders and pebbles down to grains of sand and microscopic particles of silt and mud which at some time in the past were eroded from the crust and were transported usually by rivers into the oceans and later deposited to become new rocks.

Opposite: The geological column of accepted dates for sedimentary rocks forming most of the Earth's crust. Notice the slow rate of sedimentation (averaging 0.2 millimetres per year). The column is claimed to be dated by radioactive methods but most sedimentary rocks do not contain radioactive minerals. *(Natural History Museum, London)*

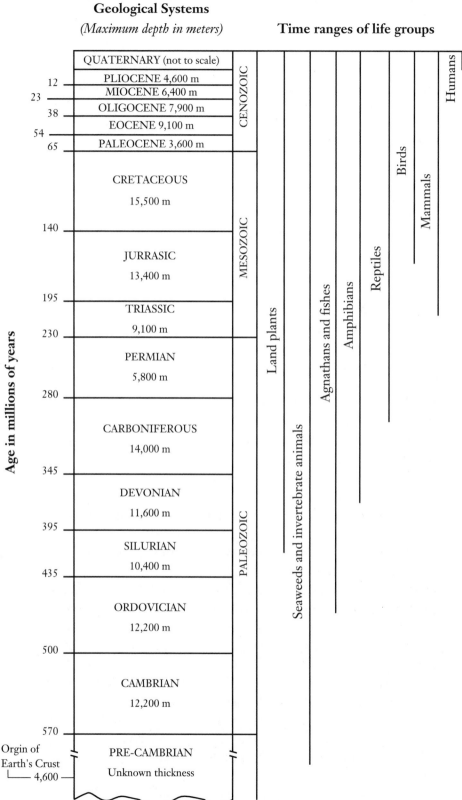

Geological Systems

(Maximum depth in meters)

Time ranges of life groups

Age in millions of years	Geological Systems	Era	
	QUATERNARY (not to scale)	CENOZOIC	
12	PLIOCENE 4,600 m		
23	MIOCENE 6,400 m		
38	OLIGOCENE 7,900 m		
54	EOCENE 9,100 m		
65	PALEOCENE 3,600 m		
140	CRETACEOUS 15,500 m	MESOZOIC	
195	JURRASIC 13,400 m		
230	TRIASSIC 9,100 m		
280	PERMIAN 5,800 m	PALEOZOIC	
345	CARBONIFEROUS 14,000 m		
395	DEVONIAN 11,600 m		
435	SILURIAN 10,400 m		
500	ORDOVICIAN 12,200 m		
570	CAMBRIAN 12,200 m		
Orgin of Earth's Crust └─ 4,600 ─	PRE-CAMBRIAN Unknown thickness		

Land plants

Agnathans and fishes

Amphibians

Reptiles

Mammals

Birds

Humans

Seaweeds and invertebrate animals

The types of sedimentary rock these particles turn into depend on the kind of material from which they are made. Sand particles are compacted to become sandstones. Silt and mud form fine-grained rocks such as shale or mudstone. Limestone, however, is not composed of particles of primary rock. Chalk, for instance, is a soft limestone said to be formed from a whitish mud or silt of organic origin.

While they were being deposited under water, these sediments often carried with them various kinds of debris, including the remains of animals and plants. The hard parts of these inclusions (such as bones, shells, and teeth) often survive as fossils, sometimes being chemically altered to become stonelike.

The extent to which the remains were preserved by burial varies greatly, especially in the case of the soft fleshy parts which usually do not survive at all. In a few instances, though—such as the trilobites preserved in very fine-grained limestones—the detail "recorded" in stone is almost miraculous and includes the microscopic crystalline structure of the eye. In rare cases, such as those of the famous Burgess shales of Canada, even soft-bodied animals are preserved as imprints in the rock.[1]

The various sedimentary rock strata are piled one on top of the other in chronological sequence, apparently representing successive episodes or phases of deposition of sediment. These strata have been examined in great detail, extensively classified, and correlated with some precision all over the country and over the world. The chalk exposed in the sea cliffs of England, for instance, can be found across much of Northern Europe, from France to as far north as Denmark. It is said to have been deposited at the bottom of a shallow sea—called the Cenomanian Sea from the Roman name for the French town of Le Mans—which covered much of Europe.

The study and interpretation of this sequence of sediments (the science of stratigraphy) is complicated by the fact that some of the beds have been laid down, only to be eroded again, giving rise to gaps in the sequence. As well, the Earth's crust has been much distorted by folding and volcanic activity. What this means is that nowhere in the world is there known to be a complete sequence of sediments, from the oldest to the most recent, so that stratigraphy is largely a matter of comparison of one outcrop with another followed by inferences as to their relationship to each other and similar outcrops. Sometimes this process is relatively simple, as in the case of the chalk cliffs which are cut through by the English Channel but which crop out in the same way, with the same fossils, on the coast of France. Sometimes, the comparisons are much more difficult and may depend on complex and sensitive techniques like finding thin strata with a characteristic electrical resistivity or characteristic mineral content which can be traced from one country to another—or even one continent to another.

The most important technique used by geologists to study the sequence of strata relies on the observation that many of the fossils contained in them appear to be restricted to one or a few particular sediments—or even a narrow band or "horizon" within a sediment. For example, the chalk cliffs of Dover are characterized by certain species of fossil sea urchin, found in the chalk but not found elsewhere. Similarly the Oxford clay—widely dug in the midlands of England to make bricks—is characterized by the fossilized shells of extinct shellfish related to the squid and known as ammonites.

Fossil species of this kind, which are believed to be associated uniquely with one type of sedimentary rock stratum, are known as zone index fossils and are used to identify that rock stratum whenever it is encountered. An example of the geological use of this technique is in drilling core samples from the seabed when prospecting for oil or natural gas. If the core sample brought up contains remains of the sea urchin *Micraster* then the geologist knows he is unlikely to strike oil in that stratum because it is a zone index fossil from the chalk, which contains no oil.

Using these principles, uniformitarian geologists have constructed the geological column—a hypothetical sequence of all known sedimentary rocks from the earliest to the most recent, each sediment correlated with distinctive fossil remains that are deemed to illustrate the animal and plant life contemporary with each phase of sedimentation. The geological column is thus considered to show not only the "record of the rocks" but also the "fossil record"—the record of life on Earth from its beginnings to the present.

It is here that the use of the geological column as a metaphor for Darwinian evolutionary processes comes in. It is clearly of the greatest practical utility to be able to identify any given geological horizon by identifying the fossils it contains. Construction engineers building cuttings for roads and railways, petroleum geologists, and many others employ this useful technique on a daily basis. It is of the greatest value to conjecture a "perfect" chronological sequence of sediments. But once the animal remains associated with each deposit are piled one on top of the other, and once relative dates are assigned to those deposits, then the fossils in them cease to be dumb, inanimate signposts and spring back to life as a *living succession* of related life forms, which evolve through the millennia that the deposition of the "record of the rocks" is supposed to have occupied. We thus have two claims that must be tested: whether the geological column is a record of processes taking millennia to unfold; and whether the fossils it contains are a living succession.

The basis of the uniformitarian view of history is encapsulated in Archibald Geikie's phrase "The present is the key to the past." What is meant by this phrase is that it is unnecessary to invoke theories of catastrophic events in the geological past in order to account for the geological succession. Instead, say uniformitarians, all the sedimentary rocks of the geological column can be explained by the same sort of events observed on the sea bottom today, working over immensely long stretches of time.

Most non-geologists (and perhaps even some geologists) will be surprised to learn that observations of modern geological processes show, however, that nowhere today are there rocks being formed anything like those in the geological column.

The main types of sedimentary rock, and those most commonly found in the geological column, are sandstones, limestones, dolomites, siltstones, mudstones, shales, conglomerates, and evaporites. Sandstones are formed from loose sands, such as those found on many beaches today, being transported and deposited by moving water. This process can be observed today, but the sands only become lithified—form solid stone—under special conditions. The chief requirement for the production of sandstone from waterborne sand grains is the presence of a cementing agent (that is, the presence of very fine grains which bind the sand grains together). This process is very familiar to anyone who has ever mixed up a batch of sand, cement, and water to tackle some job around the house—and obviously involves lithification of sand in a matter of hours, rather than millions of years. (Incidentally, the process works just as rapidly even under water, as in the case of concrete marine jetties poured between tides.)

Much the same observations apply to the lithification of shales, clays, siltstones, and mudstones. They, too, require a cementing agent in the form of smaller particles to bind them, and when such a binding agent is present they "set" rapidly.

But it is when we closely examine the other main kinds of sediments—limestones and evaporites—that a truly illuminating picture of sedimentation emerges. One of the most widely distributed sedimentary rocks in the world is limestone (and its geological cousin dolomite). Limestones are found on every continent and are composed chemically of calcium carbonate (or lime) while dolomite has a similar composition but also contains magnesium. On the face of it, limestones and dolomites might constitute perfect candidates for uniformitarian patterns of formation. Practically all of the billions of marine organisms with shells (shellfish, corals, etc.) secrete the minerals calcite and aragonite—which are chemically composed of calcium carbonate—and calcite makes an excellent cementing agent. Therefore it seems perfectly possible that limestone rocks might be forming on the sea bottom today. Indeed, until recently, uniformitarians pointed to the foraminiferal ooze currently forming on the Bahama Banks of the Atlantic as a sedimentary limestone, essentially similar to chalk, in the making. In fact the similarities that exist are purely superficial, and arose in the first place because of poor observation of the composition of chalk and modern sediments.

According to the Institute of Geological Sciences' (IGS) memoir on the Wealden district,

> Chalk is a limestone consisting of over 95 per cent of calcium carbonate. It was formerly considered to be made up almost entirely of whole and fragmented microscopic fossils, and thus comparable to the deep-sea foraminiferal oozes forming at the present day on the floor of the Atlantic. In fact the proportion of microscopic animals, chiefly foraminifera, never exceed 5 to 10 per cent of the rock. Other

investigators held the view that chalk might be a chemical precipitate or that bacterial action was involved in its formation. However, recent examination by Mr. M. Black of chalk specimens under the electron microscope has shown that the calcareous particles are calcite of organic origin . . . Coccoliths [microscopic calcareous bodies produced by planktonic algae] are present in vast numbers down to individual component crystals. The proportion of fine to coarse material varies considerably within certain limits, giving rise to chalks of different lithological character . . . Modern precipitated oozes such as those forming on the Bahama Banks, are composed almost entirely of minute aragonite crystals with a negligible proportion of coccolith material and relatively little shell debris.[2]

This issue presents problems for uniformitarian geologists in two ways. The first is that there is no sediment known to be forming on the modern sea bottom that compares with the composition of historical chalk. The second is that the exact origin of the aragonite and calcite crystals that compose the chalk of the geological column remains controversial. Uniformitarians have hypothesized that the crystals originated organically from plankton but this is not entirely supported by observation. In fact the main reason for the existence of this hypothesis is that it is necessary to support the uniformitarian view, because the only tenable alternative hypothesis is a catastrophic model of the type referred to by the IGS memoir.

If the material comprising the chalk did not have an organic origin, then it must have precipitated out of the sea water itself, and this would require sudden and cataclysmic changes in the temperature and acid-alkaline balance of huge areas of chemically saturated sea water. Nothing of this kind can be observed anywhere today. Equally important, the thickness and extent of limestone sediments in the geological column point to precipitation on a gigantic scale over huge areas (thousands of square miles) and this, too, cannot be found taking place at present.

Limestone's cousin dolomite is equally puzzling. Dunbar and Rodgers in their *Principles of Stratigraphy* say, "Although dolostone is by no means uncommon among the sedimentary rocks of the geologic record, its origin is uncertain. Probably the chief reason for this uncertainty is that, unlike other major types of sediments, it is nowhere known to be forming today, and therefore the present fails us as a key to the past."[3]

One further stratigraphically important type of sediment provides perhaps the most striking example of present processes failing to explain the past, the case of so-called "evaporites." Extensive beds of common salt, gypsum, and anhydrite occur on most continents and have been mined for centuries because of their economic usefulness. Examples include the salt deposits in Chile and Germany, and those in the northern counties of England. Again, on the face of it, such beds seem to provide excellent evidence for uniformitarian processes of slow gradual evaporation of saline lakes or seas, which clearly require great stretches of time.

The trouble with this idea is that no modern sea or lake is presently forming evaporite beds in any way comparable to these geological deposits, which are of immense thickness and great chemical purity. Another factor is that the evaporite beds contain no organic remains—no fossils—although they do sometimes contain mineral ores.

It is hardly surprising that nothing of equivalent thickness is currently forming. The salt content of the seas is the same worldwide at around 32 grams per liter.[4] To deposit even a one-meter thickness of salt over an area of only one kilometer would require the evaporation of many billions of tons of sea water. To deposit the 1,100 meters of salts in the Stassfurt deposits on the North German plain (equivalent in height to Mount Snowdon) would require the evaporation of millions of billions of tons—an ocean full of water.

The purity of these deposits and the absence of material derived from surrounding land point to them having come about not through evaporation (which is a term inspired by the uniformitarian viewpoint) but through precipitation from chemically saturated waters, in much the same manner as limestones. This idea is rejected by uniformitarians because again it implies a catastrophic origin and singular or rare events.

Russian geologist V. I. Sozansky has suggested another mechanism to account for some of these beds, based on their mineral content and lack of fossils. His proposal relates to what are called diapirs—mushrooms or domes which occur on the sea bed when softer rock beneath intrudes upward into surrounding beds. These intrusions are often accompanied by salt domes and in some cases are thought to be caused by the salt itself. Sozansky says,

> The absence of remains of marine organisms in ancient salts indicates that the formation of the salt-bearing sections was not related to the evaporation of marine water in epicontinental seas. . . . The analysis of recent geologic data, including data on the diapirs found in ocean deeps, permits the conclusion that these salts are of juvenile origin—that they emerged from great depths along faults during tectonic movements. This process is often accompanied by the discharge of basin magmas.[5]

Omer Roup, writing in the *Bulletin of American Petroleum Geologists* says,

> It is well known that salts are chemically pure formations which are void of the remains of marine organisms. If salt bearing sections were formed in lagoons or marginal seas by the evaporation of seawater, then organic matter, chiefly plankton, would have to enter the salt forming basin together with the waters. As a result the bottom sediments would be rich in organic matter.[6]

So, far from being evidence for a uniformitarian origin of sediments, the salt beds actually constitute possible evidence for a catastrophic model of sediment formation. Perhaps instead of "evaporites" we should follow uniformitarian precedent and adopt the name "catastrophites"!

But it is not merely the formation of the individual beds that comprise the geological column that does not fit with uniformitarian concepts; it is the stratification of the entire column itself that is in doubt.

All of the sedimentary rocks laid down during the Earth's history are found in clear-cut strata, one on top of another like a pile of carpets, with well-defined lines of demarcation between them. The classic uniformitarian interpretation of this finding is that the layers are successive episodes in a time sequence, that each layer is younger than those underneath it and that the cracks or joints between layers represent pauses in sedimentation. This has been the central belief of the Earth sciences since it was enunciated by Charles Lyell in 1833.

Since 1985 French geologist Guy Berthault has carried out a series of laboratory experiments involving pouring sediments into large tanks of moving water to study the internal structure of the strata, and how lamination takes place.

Berthault started his research at the Institut de Mechanique des Fluides at Marseilles and was later invited to complete his work at the hydraulics laboratory of Colorado University's Engineering Research Center.

Samples of laminated rocks were crumbled to reduce them to their original constituent particles of varying size. The particles were sorted (and colored to make them easier to identify). They were then mixed together again and allowed to flow into a tank, first in a dry state, and later into water.

What Berthault found was that when the sediments settled on the bottom they recreated the appearance of the original rocks from which they had come. But the strata were not formed by the deposition of a succession of layers as had been formerly assumed. Instead, the sediments settled on the bottom more or less immediately, but the fine particles were separated from larger particles by current flow, giving the appearance of layers.

Moreover, the lamination was found to have a thickness that was independent of the length of time taken to deposit that sediment— another fundamental assumption of classic geology. "It follows," observed Berthault, "that no deduction of the duration of sedimentation can be made by simple observation of rock laminae."[7]

The results were published by the French Academy of Sciences in 1986 and 1988 and were presented to the National Congress of Sedimentologists at Brest in 1991. Berthault pointed out that "the laminations could be shown to be caused by variations in current speed. The layer on the bottom was not laid down first and then followed by the next highest layer and so on, as required by the evolutionary interpretation of the geological column. On the contrary, the laminated layers were forming upstream sightly earlier than the lowest layers downstream."

The laboratory work has not been carried out in isolation but has been supplemented by field observations from natural disasters such as the Colorado "Bijou Creek" flood of 1965, the formation of sediments following the Mount St. Helen's eruption in 1980, and ocean drilling by the Glomar Challenger survey vessel in 1975.

What conclusions can we draw? According to Berthault, "These experiments contradict the idea of the slow build up of one layer followed by another. The time scale is reduced from hundreds of millions of years to one or more cataclysms producing almost instantaneous laminae."

These innocent-sounding words are the death knell of the idea that the existence of thousands of meters of sediments is by itself evidence for a great age for the Earth.

There is one final observation that can be made about *all* the sediments of the geological column in relation to present-day processes, and it is the greatest anomaly of all. Today there are no known *fossiliferous* rocks forming anywhere in the world. There is no shortage of organic remains, no lack of quiet sedimentary marine environments. Indeed there are the bones and shells of millions of creatures available on land and sea. But nowhere are these becoming slowly buried in sediments and lithified. They are simply being eroded by wind, tide, weather, and predators.

No carcass goes unnoticed by other animals either on land or in the sea. On the contrary, many species are carrion feeders or scavengers who specialize in seeking out and harvesting such food. Velikovsky pointed out that on the Great Plains of the United States, many millions of buffalo were slaughtered in a relatively short space of time (it has been estimated that there were 60 million buffalo when the Europeans arrived). Yet today there is not a trace of them. There are no "buffalo beds" forming on the Great Plains.

This finding is hardly surprising when one considers the conditions that must exist in order for any dead creature to become fossilized. First, and most important, it must be rapidly buried with sediment to prevent decay by bacteria or assault by predators, wave action, or weather. This sediment must be of considerable depth—

certainly inches or even feet—to prevent the remains from simply being dispersed by natural processes. Not even the most dedicated Darwinist could believe that the average rate of sedimentation of the uniformitarian geological column (0.2 millimeters per year) is capable of providing such rapid burial.

The detail and completeness with which many fossil specimens are preserved (the eyes of the trilobite, the scales of fish and even the skin of dinosaurs) is a clear indication that the creatures were rapidly buried under considerable depths of sediment. The very size of some specimens, such as the larger land-living dinosaurs makes it absurd to suppose that they could have been preserved in a few millimeters of sediment. To preserve by burial an adult brontosaurus, or diplodocus, would require tens of meters of sediment, and these quantities can only be explained by catastrophic events, rather than uniform deposition.

Before leaving the metaphorical world of the uniformitarian geological column, and turning to a detailed examination of the fossils it contains, there is one more important sediment to examine: one about which there have been many curious and remarkable findings, but which still holds many mysteries—coal.

CHAPTER 8

An Element of Unreality

C OAL HAS THE UNIQUE DISTINCTION OF BEING BOTH the best known sedimentary rock in the geological column and the best known fossil. It is a carbon-rich rock that is formed from the decomposition of great masses of vegetation—typically whole forests. But although it is such a familiar everyday material and despite being on the syllabus of every elementary school, the real story of coal is a mystery, with not one but many twists in the tale.

Coal varies in density and color and is classified according to its carbon content. A low carbon coal is called lignite. A little more carbon ranks it as bituminous coal. A high carbon content material is classified as anthracite.

According to uniformitarian geologists, coal is formed over the customary many millions of years. Masses of vegetation, usually great forested plains, are said to have become swampy, and formed peat bogs. These peat bogs have later been inundated by the sea and then have been crushed by substantial depths of marine sediments which are laid down on top of them, converting the peat into coal.

The coalification process envisaged by uniformitarians has two stages. The first is analogous to that taking place in a garden compost heap—the relatively rapid decomposition of the vegetable material by bacterial and fungal action to form peat, and then the long slow compression by overlying rocks. The longer the pressure is sustained, the higher the carbon content of the coal, and the higher its rank.

To begin with, pressure is applied to the decomposing peat bog and it turns into lignite: brown in color, soft, and burning with little heat when lit. More pressure and a few million years, and the lignite turns into bituminous coal: harder than lignite, darker in color and giving off much more heat when burned. After many more millions of years, the bituminous coal has been converted to anthracite: jet black, hard, and burning fiercely when lit.

The geological events that gave rise to the formation of the largest coal deposits are said to have begun in a remote period named the Carboniferous, after its most characteristic product, which began 325 million years ago and lasted for 45 million years. The sediments laid down in this period are referred to as the coal measures and they are represented in many countries around the world.

The coal measures are immensely thick sedimentary deposits containing a variety of rock types, occurring in sequences which are often repeated. Typically, these sequences include beds of shale containing freshwater fossils; overlain by strata of coal; overlain in turn by thick beds of limestone containing fossils of marine animals. These repeated sequences, called cyclothems, are a key feature of the Carboniferous rocks and are always associated with coal deposits.

There are a number of important factors to notice about the uniformitarian scenario. A two-stage process is necessary to the theory in order to provide the initial high temperature which creates the lowest rank of coal or lignin. It is necessary to the theory because the only other mechanism uniformitarians have available to accomplish coalification is a very slow rate of deposition of the overlying rock—typically 0.2 millimeters per year. Clearly this alone would not account for the fossilization of a forest. On the contrary, such low uniform rates of deposition would simply allow the trees to rot away and be dispersed by wave or current action.

A second necessary part of the theory is that the forested land must be inundated by the sea. This is necessary first because the sediments overlying coal strata contain marine fossils, but more significantly because uniformitarians need the pressure of accumulating overlying rocks to account for the slow transformation of peat into coal. For these reasons, uniformitarians suggest that coal-forming forests were on low-lying coastal plains or basins (usually in swampy conditions) which were subject to periodic marine invasion. This in turn means that there has to be periodic sinking or subsidence of the basin in which the forest grows, and this is said to be caused by major movements of the Earth's crust.

To summarize the uniformitarian coalification process: a forest grows up in a basin or plain beside the sea. The forest becomes swampy, but with fresh water. A vast peat bog forms. The Earth's crust shifts, the basin or plain sinks and the sea covers the peat bog. Over millions of years, limestone sediments are laid down on the bottom of the sea, compressing the peat bog and increasing the rank of coal thus formed. At the end of this period, the land rises and the basin or plain is exposed once more. Again, a forest springs up on the reclaimed land; the forest becomes swampy with fresh water; a vast peat bog forms. The Earth's crust shifts and the plain or basin sinks once again beneath the sea. More marine limestones are deposited, and so on.

If uniformitarians claimed that this had happened once, twice, or even three times in the same spot, we would have to grant that their scenario could have occurred. But in the coal measures this sequence is repeated not two or three times, but as much as *sixty* times.

According to Hollingsworth, writing on coal formation,

> In the case of the permo-Carboniferous [rocks] of India, the Barakar beds of the Damuda series, overlying the Tachir boulder bed, includes numerous coal seams, some up to 100 feet thick, occurring in a well-developed and oft-repeated cycle of sandstone, shale, coal . . . the vegetation is considered to be drift accumulation. The concept of periodic epirogeny is a reasonable one, but a more or less complete cessation of clastic [derived rock particle] sedimentation in the lacustrine basin during coal accumulation is difficult to account for on a wholly diastrophic origin. As an explanation for the fifty to sixty cycles of the Damuda system, it has an element of unreality.[1]

This "element of unreality" also attaches to some other aspects of the uniformitarian view of coal origins. In 1945 Melvin Cook was appointed by the U.S. Navy to direct a high explosives group. Part of the group's work was to develop safer explosives for use in coal mining and as part of this project he made a special study of the occurrence and characteristics of coal. Cook points out that wood is composed mainly of cellulose with about one quarter being a material called lignin. This chemical composition is the basis of wood's (and coal's) usefulness as a heat source. Burning wood can give off almost as much heat on a pound-for-pound basis as a powerful explosive like TNT. According to Cook, the dehydration decomposition of wood gives rise to exothermal heat in the range 400 to 800 calories per gram, compared with TNT which generates about 1,000 calories per gram. If wood is compacted and put under pressure (as by being buried) the decomposition initiated by pressure alone will supply the necessary temperature to convert wood to coal, making the hypothetical biochemical or peat bog stage quite unnecessary.[2]

The best evidence that pressure, rather than time, is the cause of coalification comes from examining the rank of coal in relation to the depth of its deposit. In the United States, the Pittsburgh coal seam runs between Ohio and Pittsburgh and the strata in which the seam is contained dip downwards into the Earth at the rate of 20 to 40 feet per mile, with the coal at the easternmost end of the seam several thousand feet deeper than at the western end. As the seam goes deeper, the grade of coal increases: the deeper the burial and the greater the compression of overlying beds, the further the process of coalification has proceeded. In this case, the reaction would be started without any microbiological attack and could be achieved rapidly by pressure alone.

If coal was formed relatively quickly by rapid burial under marine sediments, then swamps, peat bogs, microbiological attack and millions of years of gradual deposition and slow pressure are no longer needed. Once again, as with the other sediments of the geological column, the key question is not so much how they were formed, but how quickly they were formed.

There is so much evidence on this point that it is hard to see how it could have been overlooked by uniformitarians. Although fossils are relatively rare in the coal itself, it is common for miners to discover large inclusions in the coal seams, such as boulders. And when fossils are found, they can be spectacular.

In 1959 Broadhurst and Magraw described a fossilized tree, in the position of growth, from the coal measures at Blackrod, near Wigan in Lancashire:

> The tree was preserved in the form of a cast and the evidence suggested that the cast was at least 38 feet in height. The original tree must have been surrounded and buried by sediment which was compacted before the bulk of the tree decomposed so that the cavity vacated by the trunk could be occupied by new sediment which formed the cast. This implies a rapid rate of sedimentation around the original tree.[3]

Broadhurst also says that such fossil trees in position of growth are far from rare in Lancashire and points out that in 1956 Teichmuller reached the same conclusions for similar trees in the Rhein–Westfalen coal measures of Germany.

In 1878 miners at Bernissart, a small village in the Mons coalfield of Southwest Belgium, made a spectacular discovery when they uncovered a fissure in the coal seam packed full of intact dinosaur skeletons, at a depth of 322 meters. Thirty-nine skeletons of the dinosaur iguanodon were recovered, from a fissure 100 feet high, many of them complete, and are now on display in the Royal Institute of Natural Sciences in Brussels.

The most striking thing about these creatures is that they measured 10 meters in length, stood several meters high, and weighed in the region of 2 tons apiece. For their bodies to be rapidly buried would require rates of deposition thousands or even millions of times greater than the average 0.2 millimeters per year proposed by uniformitarians.

Geologists and paleontologists are well aware from their field studies that the same thing applies to all of the sediments of the geological column. It is commonplace to find large fossils in position of growth or taking up their original volume, such as horsetails in Deltaic sandstones, corals in the Oolitic limestones, giant ammonites in the Portland beds, and tree trunks in strata of many kinds.

Sedimentary rocks are said by uniformitarians to take millions of years to form at slow rates of deposition. But full-size trees found in position of growth point to rapid burial. This fossil tree excavated from Carboniferous rocks near Edinburgh stands on the grounds of the Natural History Museum, London. *(Photo: author)*

Certainly the custodians of the geological column at London's Museum of Natural History cannot fail to be aware of such discoveries. Each morning, on their way to work, they cross the museum's grounds and pass a fossilized tree trunk from the Lower Carboniferous, excavated at Craigleith quarry, Edinburgh, which originally measured some twenty or more feet in height.

If, as the evidence presented here suggests, coal was formed rapidly, what about the other sedimentary rocks in the geological column? Might they, too, have been formed in some relatively rapid catastrophic processes rather than slowly over millions of years?

CHAPTER 9

When Worlds Collide

Putting to death the bearer of bad tidings is an activity usually associated with the more uninhibited Roman emperors or Eastern despots. Yet messenger-shooting can be just as common in scientific and academic circles when the bad news concerns one of science's sacred cows.

One messenger who published findings that challenged the received wisdom on geological history—and who was mugged by his fellow scientists for his trouble—was Immanuel Velikovsky, the American psychologist whose 1950 book *Worlds in Collision* caused a virtual panic in the academic community. Velikovsky proposed that a near collision between the Earth and other planets of the solar system caused catastrophic geological events in the Earth's history and that most of the major features of the Earth's crust are testimony to these events. He also sought to establish a short time scale for the Earth's history.[1]

Velikovsky's book caused the sort of reaction among the scientific fraternity that one might expect had he proposed the collision were actually going to happen next Friday. His treatment was so shameful that it led professor Alfred De Grazia of New York University, writing in the journal *American Behavioral Scientist*, to observe that Velikovsky's book:

Gave rise to a controversy in scientific and intellectual

circles about scientific theories and the sociology of science. Dr. Velikovsky's historical and cosmological concepts, bolstered by his acknowledged scholarship, constituted a formidable assault on certain established theories of astronomy, geology and historical biology, and on the heroes of those sciences. Newton himself, and Darwin were being challenged, and indeed the general orthodoxy of an ordered universe.

What must be called the scientific establishment rose in arms, not only against the new Velikovsky theories but against the man himself. Efforts were made to block the dissemination of Dr Velikovsky's ideas, and even to punish supporters of his investigations. Universities, scientific societies, publishing houses, the popular press were approached and threatened; social pressures and professional sanctions were invoked to control public opinion. There can be little doubt that in a totalitarian society, not only would Dr Velikovsky's reputation have been at stake, but also his right to pursue his enquiry, and perhaps his personal safety.

As it was, the "establishment" succeeded in building a wall of unfavorable sentiment around him: to thousands of scholars the name Velikovsky bears the taint of fantasy, science-fiction and publicity.[2]

Geologists and astronomers were so virulently opposed to Velikovsky's book that they threatened to boycott the scientific textbooks of his publisher, Macmillan, forcing the firm to turn Velikovsky's work over to another publisher, Doubleday, who was not involved in textbook publishing and hence not susceptible to academic blackmail. Today, only forty years later, a concept closely similar to Velikovsky's is widely accepted by many geologists—that the major extinction at the end of the Cretaceous (and possibly other extinctions) were caused by collision with a giant meteor or even asteroid.

Velikovsky was treated so badly by the scientific community that he determined to back up his theory with an unchallengeable body of evidence. He spent five years researching a second book on the catastrophist theme, *Earth in Upheaval*, in which he provides detailed evidence on scores of geological structures and paleontological finds which are inexplicable on any basis other than a catastrophic origin. Moreover, the extent of the catastrophe required to produce these structures he showed to be global, and the energies needed on a cataclysmic scale.[3]

I will not needlessly repeat Velikovsky's very detailed research (his books are listed in the bibliography) but I will summarize three of his examples which are not only well attested to by multiple sources, but which cannot be accounted for on any but a catastrophist model. They are the young age and rapid building of the world's mountain chains; the gigantic extent of certain rock formations, requiring singular, acute causes; and the occurrence of extinctions on a massive scale. There are also two mysteries which require explanation: anomalies relating to glaciation in the Ice Age, and the existence of beds of fossils thousands of feet deep but containing the remains of terrestrial, rather than marine, animals.

The major mountain chains are conventionally believed to be the result of pressure at the edges of the continental "plates": in effect they are the buckling of the edge of one plate by another, rather like two cars in a road accident. The Andes in South America and their North American counterpart, the Rockies, are said to be caused by pressure from the Pacific plate on the American plate. True to the uniformitarian model, this movement of plates and consequent mountain-building is deemed to have taken place not at all like a traffic accident, but very slowly over millions of years at a rate in the order of 1 to 10 centimeters per year. The trouble with this idea is that there is a substantial body of evidence pointing to rapid mountain-building occurring in the recent past, thousands rather than millions of years ago.

In the Alps, for example, there are numerous sites of human occupation at altitudes that must be far above their original level. Human artifacts dating from the Pleistocene or Ice Age have been discovered in caverns at Wildkirchli, near the top of Ebenalp, at 4,900 feet (nearly 1 mile) above sea level. Even more astonishing is the cavern of Drachenloch near the top of Drachenberg, south of Ragaz, which was also occupied by humans during the Pleistocene and is some 8,000 feet above sea level (well over a mile and a half high). There are other examples in other continents of routine human habitation at extraordinary heights, especially in the Andes. This appears to point decisively to a substantial part of mountain-building activity taking place in the recent past.

A study of the Ice Age in India by Helmut de Terra of the Carnegie Institute and Professor T. T. Paterson of Harvard University concluded that the Himalayas were still being built during the Ice Age, and reached their present great height only during the historical era. "Tilting of terraces and lacustrine beds," wrote de Terra in *Studies on the Ice Age in India and Associated Human Cultures* in 1939, indicates a "continued uplift of the entire Himalayan tract" during the last phases of the Ice Age.

At 12,500 feet up in the Andes (two and a half miles above sea level) is the deserted but well-preserved city of Tiahuanacu. It is in a region where corn will not ripen and its altitude is too high today to support life for anyone other than a tribe of mountaineers. In 1910 the president of the Royal Geographical Society, Leonard Darwin, suggested that the mountains had risen considerably after the city was built, and it is hard to find an alternative explanation that is credible. If the Andes were as little as 3,000 feet lower, corn would ripen in the basin of Lake Titicaca and the site of Tiahuanacu would support a sizeable population.

The second indicator of catastrophism on a grand scale is the extent of certain geological formations, principally volcanic lava flows. In North America an area of 200,000 square miles in Idaho, Washington State, and Oregon, known as the Columbia Plateau, is covered by lava to a depth as great as 5,000 feet (almost 1 mile). Uniformitarianism could never account for such beds. This quantity of lava exceeds by many orders of magnitude all the lava flows from all the world's currently active volcanoes. And there are similar deposits on other continents, such as the Deccan traps in India, 250,000 miles square and several thousand feet deep, the lava bed of the Pacific Ocean and the lava dykes that cross South Africa.

The third indicator of historical catastrophes is that of extinctions on a huge scale. A common rock in the geological record is the Old Red Sandstone. The northern half of Scotland from Loch Ness to the Orkneys exposes this rock formation in myriad sites to a total depth of more than 8,000 feet (twice the height of Ben

Nevis). In an area 100 miles across, the Old Red Sandstone contains the fossils of billions of fish, contorted and contracted as though in convulsion and resulting apparently from some catastrophic event.

Describing the fossil fauna in his 1841 study, *The Old red Sandstone*, Hugh Miller wrote, "Some terrible catastrophe involved in sudden destruction the fish of an area at least a hundred miles from boundary to boundary, perhaps much more. The same platform in Orkney as at Cromarty is strewed thick with remains, which exhibit unequivocally the marks of violent death." The same scene is found at Monte Bolca in northern Italy where Buckland, writing in 1836, observed, "The circumstances under which the fossil fishes are found at Monte Bolca seem to indicate that they perished suddenly. The skeletons of these fish lie parallel to the laminae of the strata of the calcareous slate; they are always entire and closely packed on one another. . . . All these fishes must have died suddenly."

Similar formations are found in the coal measures of Saarbrucken on the Saar, the calcareous slate of Solenhofen, the blue slate of Glaris, and the marlstone of Oensingen in Switzerland and of Aix-en-Provence. In the United States there are comparable formations such as the black limestones of Ohio and Michigan, the Green River bed of Arizona, and the diatom beds of Lompoc, California. D. S. Jordan reported finding in the Monterey shale of California enormous numbers of the fossil herring *Xyne grex*. Jordan estimated that more than 1 billion fish, averaging 6 to 8 inches in length, died on 4 square miles of sea bed.[4] Ladd points out that catastrophic death of fish on a large scale does occur sometimes today, in the case of so-called "red water" for example. What does not occur, however, is death on a scale of billions. Nor do the victims become rapidly buried in thousands of feet of sediment and fossilized—their carcasses are preyed on by scavengers.

The two mysteries that have a bearing on catastrophes receive little publicity, yet are tantalizing in the extreme, crying out for an answer. The first has to do with glaciation during the Ice Age. In the 1830s a Swiss naturalist, Louis Agassiz, realized that much of Europe must once have been covered in ice. Agassiz became fascinated by glaciers in his native Switzerland. He even built a hut on a glacier at Aar and lived in it so he could study the movement of the ice front. He deduced from this study all the actions of which glaciers are capable: transporting large quantities of rocks and stones (including huge boulders) which will be left behind when the ice melts (moraines); gouging out U-shaped valleys; and cutting striations in the underlying rock surface.

Agassiz converted Dean Buckland, influential president of the Geological Society, to belief in an Ice Age by showing him distinctive glacial features in Scotland, and he converted Charles Lyell by showing him some moraines within two miles of his father's house. Having secured such powerful backing, Agassiz's theory was certain of universal acceptance. A key consequence of this widespread acceptance has been the tendency to ascribe *all* inexplicable terrestrial features to glacial action. This action, of course, is believed to have taken place according to the uniformitarian model over hundreds of thousands or even millions of years. One particular feature which glaciation is used to explain is the occurrence of "erratics"—substantial rocks which are geologically out of place. On the coast of Scotland are large quantities of rocks which have been transported from the mountains of Norway. In North America erratic blocks of Canadian granite are found over ten of the northern United States from Maine to Ohio. All these are believed to have been moved by glaciers working slowly but surely.

And this is where the puzzle comes in. In Eastern Europe there are many erratics strewn across the Russian plains. But whereas in Finland and the northern provinces, these blocks are large, they get uniformly smaller as one goes south. A similar pattern of uniform grading of erratics is found elsewhere in Europe and North America. This distribution points not to ice but to water action, and water on a huge scale. The uniform grading also points to turbulent flood conditions gradually abating.

In addition marine fossils are found *on top* of glacial deposits as in the case of the whale skeletons found in bogs covering glacial deposits in Michigan. According to Dunbar in *Historical Geology*, whale fossils have also been found 440 feet above sea level north of Lake Ontario; more than 500 feet above sea level in Vermont; and some 600 feet above sea level in the Montreal area. As Velikovsky observes, "Although whales occasionally enter the mouth of the St. Lawrence river, they do not usually climb the surrounding hills."[5]

The second mystery is one that has intrigued many geologists since the early nineteenth century, including Alfred Russell Wallace, codiscoverer with Darwin of evolution by natural selection. The mystery concerns a range of hills called the Siwalik Hills north of the Indian capital Delhi. The hills, some 2,000 to 3,000 feet high and several hundred miles long, are actually the foothills of the Himalayas. The Siwaliks contain extraordinarily rich beds crammed with fossils: hundreds of feet of sediment, packed with the jumbled bones of scores of extinct species. Many of the creatures were remarkable; including a tortoise 20 feet long and a species of elephant with tusks 14 feet long and 3 feet in circumference. Other animals commonly found include pigs, rhinoceroses, apes, and oxen.

Most of the species whose fossils are found are today extinct, including some thirty species of elephant of which only one has survived in India. Beds of this sort are common in the geological record, as in the case of the fish beds referred to above. But the Siwalik beds contain the remains of *terrestrial* animals, not marine creatures. These animals must have been killed by some singular event over a relatively short space of time, and an event which took place on land. And whatever the nature of the event, it resulted not only in catastrophic extinction of many species but also the formation of beds of sediment thousands of feet thick.

It is sometimes suggested that the animals were killed by the onset of the Ice Age. But no mechanism has been proposed that would account for such large numbers being killed by ice creeping along at a few centimeters a year, or for their rapid burial in thousands of feet of sediment.

It was also proposed that the Siwalik deposits were alluvial and represented debris carried down by the torrential Himalayan streams. But it was realized, as Wadia wrote in his *Geology of India*, that the alluvial explanation "does not appear to be tenable on the ground of the remarkable homogeneity that the deposits possess" and their "uniformity of lithologic composition" in many different and isolated basins miles apart.[6]

Thirteen hundred miles from the Siwalik hills in central Burma are deposits of a very similar nature, containing remains of mastodon, hippopotamus, and ox along with large quantities of fossil wood—thousands of fossilized tree trunks and logs scattered in the sandstone sediments. In total the deposits may be as much as 10,000 feet thick and there are two distinct fossiliferous horizons separated by 4,000 feet of sandstone.

Velikovsky is still a favorite target for attack by uniformitarian geologists and the many examples of catastrophism he unearthed and presented are still vehemently rejected. One geologist wrote to me to claim that there is nothing unusual or unprecedented about the Siwalik hills beds, although the only example of a comparable formation he was able to offer were beds that are an order of magnitude smaller in depth, and lateral extent, and which do not contain a comparable fossil fauna.

This continued resistance is strange given that graveyards of terrestrial animals are commonplace. In his books on dinosaurs, Dr. Edwin Colbert gives numerous examples: in New Mexico "there were literally scores of skeletons on top of one another and interlaced with one another. It would appear that some local catastrophe had overtaken these dinosaurs, so that they all died together and were buried together."[7]

At Como Bluffs, Wyoming, referred to earlier, "the fossil hunters found a hillside literally covered with large fragments of dinosaur bones. . . . In short it was a veritable mine of dinosaur bones."

In Alberta, Canada, "innumerable bones and many fine skeletons of dinosaurs and other associated reptiles have been quarried from these badlands, particularly in the 15-mile stretch of river to the east of Steveville, a stretch that is a veritable dinosaurian graveyard."[8]

Colbert refers also to the Belgian dinosaur find mentioned in the last chapter: "Thus it could be seen that the fossil boneyard was evidently one of gigantic proportions, especially notable because of its vertical extension through more than a hundred feet of rock."

These examples can be multiplied almost endlessly, yet few modern geologists are prepared to accept that major features of the Earth's crust could have been caused by singular events, because such an admission seems to open the door to some kind of geological anarchy which threatens the orderly arrangement of exhibits in their glass cases.

So far this book has been concerned with the purely geological question of how, and how fast, the rocks of the geological column were formed. An even more significant question from the point of view of Darwinian evolution theory is, What conclusions about the origin of life can be drawn from the fossils contained in the geological column? To seek answers to this question we must turn from geology to its close relative, paleontology.

CHAPTER 10

The Record of the Rocks

O NE AFTERNOON IN 1822 a medical practitioner from Brighton on the English coast, Dr. Gideon Mantell, took his wife for a walk in the beautiful spot of Ashdown Forest. Mrs. Mantell must have been unusually observant for she picked up a strange tooth from a pile of road-mending stone they passed. Puzzled by the tooth, Mantell showed it to the leading geologist of his day, Charles Lyell (who in turn showed it to France's Baron Cuvier) but neither was able to identify the animal from which it had come.

Mantell turned to his own professional body, the Royal College of Surgeons, and made a systematic search through the college's collection of teeth but without success. He was about to give up when the curator showed him a newly discovered lizard specimen just arrived from America, called an iguana. By an extraordinary coincidence, its teeth were closely similar to that found by his wife, and Mantell realized he was holding the tooth of an unknown extinct reptile of great size. The tooth was described by William Conybeare, who coined the name *iguanodon* to describe its long dead owner—the first dinosaur to be identified.

The tale of this seminal event is in some ways a parable for the history of paleontology as a whole: comparative anatomy has played a decisive role in the development of the science; much can be deduced from apparently meager finds, as long as their significance is appreciated; chance plays a substantial part in paleontological discoveries; and a little intelligent guesswork can often go a long way to solving a tantalizing mystery.

But there is a negative aspect to the parable. His discovery made Gideon Mantell obsessed with finding more dinosaur remains and he filled their Brighton home with so many rock specimens that his wife left him, never to return. This obsession with fossils is by no means rare and has something in common with the gold fever experienced by prospectors. It is an obsession that has played a part in paleontology and evolution theory in many ways over the past 150 years.

The modern, stratigraphical significance of fossils came about through the large-scale engineering works undertaken at the beginning of the nineteenth century. Hundreds of miles of roads, canals, and railway cuttings were dug across Britain exposing rocks of every kind along with the fossils they contained. William Smith, the "father of English geology," was an engineer responsible for cutting canals and he noticed that similar sequences of rock strata recurred in different places and that they often contained similar fossils. Wherever his workers' picks struck the creamy rocks of the lower oolites, there he found the distinctive mollusc *Trigonia*. When they dug the tough blue shales of the Lias, there he found the oyster *Gryphaea*, called "devil's toenails" by country people.

Smith began to draw up the first geological maps—of the countryside around the city of Bath—marking the different beds of rock in different colors. Such charts were immensely useful to him in siting and digging his canals, telling where to find building stone for aqueducts and bridges; where to find clay to waterproof his canal across porous rocks; which line to take across the countryside. It was largely because of his unaided pioneering efforts in publishing the first geological map of Britain in 1815 that the Geological Survey of Great Britain was established soon afterward.

Smith made one further observation about the fossils he discovered. He realized that they were the remains of marine creatures and that the rocks he was looking at were the floor of an ancient sea—in fact a succession of such seas—which had covered Britain sometime in the past. He realized too that the succession of rocks was accompanied by a succession of fossils—the oldest at the bottom, the newest at the top.

Smith's legacy to geology is of incalculably great value. Unfortunately, when he decided to distinguish different rock formations by different colors in his first maps, he unwittingly bequeathed geology a technique that was to have unexpectedly ambiguous results later on. For he made it possible—indeed, almost inescapably natural—to associate a *chronological succession* of rocks with an *evolutionary succession* of life forms in the past.

Smith had been interested in fossils since he was a young boy living on a farm near Oxford and had amassed a large collection of fossil specimens. As a result of his research as a canal engineer, he rearranged his collection of specimens stratigraphically: all the fossils from the Bradford clay in one drawer, all those from the Lias shales in another, and so on. You might think this is an obvious thing to do, but it is more natural to arrange fossils collected from different locations in *biological* groups, for comparison—all the shark's teeth together, all the sea urchins together, and so on.

When you make such a biological arrangement, you see that a great many species are very stable in shape and size. An oyster from the Jurassic period looks very much like an oyster from any later period, including one washed down with champagne today in the Ritz. On the other hand, one of the main reasons for supposing, as Darwinists and other evolutionists do, that species evolve as you proceed up the geological column is that some types of creatures disappear and are replaced by something similar yet distinctively different. For instance, if you walk along the beach from Lyme Regis in Dorset to the neighboring town of Charmouth, you will find that the rocks in the cliffs have been tilted at an angle by earth movements. As you travel along the beach you are able to look higher and higher up the geological succession. What you find in those rocks as you pass along are different species of ammonite (a spiral shellfish related to the present-day pearly nautilus). In the lowest bed is a genus called *Asteroceras;* in the next, one called *Amaltheus;* and in the highest *Harpoceras.*

Uniformitarian geologists believe these rocks took millions of years to form. Darwinists say that the successive ammonite species represent a line of descent: that the ammonite *Harpoceras* near present-day Charmouth is the remote offspring of *Asteroceras* at the Lyme Regis end. The fact that there are gaps in the fossils and no transitional forms intermediate between the various species does not alter this conviction. Because the rocks are a succession and took millions of years to lay down, then the fossils they contain are a living succession also.

In one sense it is very surprising that uniformitarian geologists should think this way about the biological past, since it is quite contrary to that most fundamental principle of geological history: the past can be understood in terms of the present. In the animal world the most striking thing about species today is their discontinuity. The living world consists mostly of gaps between species; gaps that remain unbridgeable even in the imagination. The fossil record indicates clearly that the living world also consisted of gaps in every past age from the most recent to the most remote. Yet Darwinists believe that while the present consists of gaps, the past was a perfect continuity of evolving species—even though this continuity is not recorded in the rocks—and they have devoted immense efforts to find credible sequences of fossil ancestors and descendants.

Probably the best known of such sequences is that of early horses discovered mainly in North America. Illustrations of this sequence figure prominently in textbooks on paleontology and in natural history museums around the world. It is due to the passionate bone collecting of O. C. Marsh, professor of paleontology at Yale University, and his intense rival Edward Cope. Their materials were arranged by Henry Fairfield Osborn, director of the American Museum of Natural History and his assistant William Matthew. As early as 1874 Marsh declared that "the line of descent appears to have been direct and the remains now known supply every important form."[1]

The sequence begins with a tiny creature the size of a dog, poetically named *Eohippus* (dawn horse) by Marsh, with four toes on its front legs, three toes on its hind legs, and teeth suited to forest browsing. This creature is said by Darwinists to come from the lower Eocene, about 50 million years ago by their dating methods. In beds of Oligocene age (around 30 million years old) are found the remains of *Mesohippus*, a creature the size of a sheep, with three toes on each leg. In Miocene beds, said to be 15 million years old, are found fossils of *Merychippus*, still with three toes but

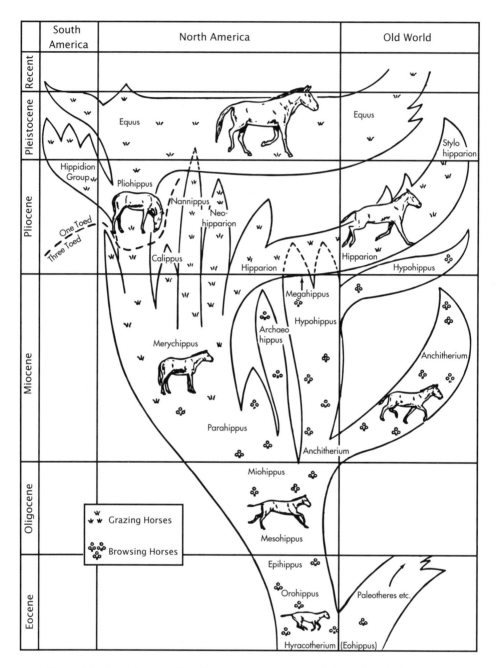

One of the best-known "evolutionary sequences," that of early horses. Although the diagram shows an unbroken line, there are major gaps in the sequence, for example between *Eohippus* and its supposed ancestor, and between *Eohippus* and its supposed descendant, *Miohippus*. (*Drawing from* Horses *by George G. Simpson*)

walking on tiptoe and with teeth adapted to grass-feeding habits. And from Pliocene beds of around 7 million years ago come the remains of *Dinohippus*, an animal the size of a small pony with only one toe (hoof) with rudiments of toes on either side, and teeth fully adapted to grazing.[2]

There is no question that these remains, when placed together, are strongly suggestive of an evolutionary development. They show what the evolutionary model predicts the fossil record should show. In fact it is comparative anatomy of this sort that provides the strongest body of evidence for evolution in the first place. And it is easy to understand the enthusiasm and speed with which the American Museum set up its display—a display rapidly copied by the British and other museums.

From a purely scientific standpoint, however, there are two difficulties with this sequence. The first is that although the fossil record has been bountiful enough to provide these intermittent remains, it has been consistently reluctant to yield up any remains that are actually transitional between them. The similarities between *Eohippus* and *Mesohippus* are great. But their differences are greater still. Bones of *Eohippus* and bones of *Mesohippus* have been found in a number of places. But bones of the animals that are said to connect them in lineal descent are not merely rare—they are nonexistent. And the same thing is true for most of the animals in the sequence: transitional species are not merely unusual they are missing entirely.

The second difficulty is that, given the continued existence of gaps in the fossil record, and the continued failure to find fossils of the hypothetical intermediate species, then to call the *Eohippus* sequence an evolutionary series is not a scientific theory—it is an act of faith, a matter of belief. It is perfectly true that an intelligent rational person can examine the remains and be convinced that they represent an evolutionary sequence, but not by virtue of any evidence that has been adduced, since the *Eohippus* sequence is not evidence for evolution. It is evidence for the former existence of different species of quadruped with a striking similarity, not evidence of a relationship between them. And it is this, the relationship—if any—which is the very matter in question.

According to Professor Garret Hardin:

There was a time when the existing fossils of the horses seemed to indicate a straight-line evolution from small to large, from dog-like to horse-like, from animals with simple grinding teeth to animals with the complicated cusps of the modern horse. It looked straight-line—like the links of a chain. But not for long. As more fossils were uncovered, the chain splayed out into the usual phylogenetic net, and it was all too apparent that evolution had not been in a straight line at all, but that (to consider size only) horses had now grown taller, now shorter with the passage of time. Unfortunately, before the picture was completely clear, an exhibit of horses as an example of orthogenesis had been set up at the American Museum of Natural History, photographed, and much reproduced in elementary textbooks. (where it is still being reproduced today).[3]

Hardin was writing in 1961 but, regrettably in light of what we now know, the same basic display is still on view at the British and other Natural History Museums, making largely the same claims, and it is still being reproduced in textbooks and the current edition of *Encyclopaedia Britannica*.

One of the principal modern champions of Osborn's evolutionary sequence for horses has been George Simpson. Simpson himself made important fossil horse discoveries in Texas in 1924, and his 1951 book *Horses* first encapsulated all the findings of the American Museum team. The book makes fascinating reading, yet its author seems unaware of the many contradictions it contains. On the general question of horse evolution, Simpson says, "The history of the horse family is still one of the clearest and most convincing for showing that organisms really have evolved," and "there really is no point nowadays in continuing to collect and to study fossils simply to determine whether or not evolution is a fact. The question has been decisively answered in the affirmative."

Compare this certitude with the following selection of quotations from the same book by Simpson (the paragraphs are quoted in sequence but are not connected in the original):

Bonediggers have not, as yet, had the good fortune to find the precise immediate ancestors of eohippus or those that would show exactly where and when the horse family first arose.

In Europe there are no really good collecting fields of early Eocene age and fossils are few, but eohippus forms a considerable percentage of those that are known. In the richer early Eocene beds of North America . . . eohippus is an abundant fossil. Hundreds . . . of specimens have been found, although most of them are fragmentary, single teeth or scraps of jaws or other bones. For some reason not clear to me, common as eohippus remains are, it is most unusual to find so much as a whole skull and skeletons anywhere near complete are exceedingly rare. As far as I know, only four skeletons have ever been reconstructed and mounted.

It happens that fossil mammals from around the very end of the Eocene and the very beginning of the Oligocene have not been well known in America. In recent years this gap in knowledge is being filled, but we still do not know enough about the animals of that important time of transition from one epoch to another. This applies also to the horses, and around this time there is a slight break in our otherwise practically continuous knowledge of horse history.

The teeth of this horse [*Epihippus*] were more progressive than those of any typically Eocene form and more primitive than any of unquestioned Oligocene age, but somewhat nearer the earlier type. The skeleton is practically unknown and we can only guess that, when discovered, it may more fully confirm the reasonable inference that American Oligocene horses were directly derived from *Epihippus*. It remains possible however, that the immediate ancestor of the Oligocene horses lived in some other region where its bones have not been found.

One other peculiar and extinct group should be mentioned . . . the pygmy horses of the Miocene. These are united under the name *Archaeohippus*. . . . It is a pity that the skeleton is so incompletely known that no mounted specimen or restoration can yet be made. At any rate this reversal of the usual, but by no means constant, tendency for horses to increase in size is of extraordinary interest.

Simpson concludes his book with what seems to me a remarkable statement of the rarity of all the finds on which the lineage of horses is based. Under the heading "Where to See Fossil Horses" he writes:

> Complete mounted skeletons of fossil horses are rarities. They are seldom found and their preparation is a long, laborious, highly skilled and expensive job. Much the greater part of the display and research materials of fossil horses consists of partial skeletons or, especially, isolated bones, skulls, jaws and lesser fragments. Of these partial fragments tens of thousands are known. Of mounted skeletons, there are fifty-odd in the United States. . . . As far as I know, there are no mounted skeletons of *Epihippus, Archaeohippus, Megahippus, Stylohipparion, Nannippus, Calippus, Onohippidium*, or *Parahipparian*, and none in the United States of *Anchitherium* or *Hipparion*.[4]

Of course, it is not essential to have a complete skeleton in order to describe an extinct creature anatomically with a reasonable level of confidence. It is, however, more than a little disturbing to learn that the descent of horses is being offered as *the* decisive evidence in favor of evolution on the strength of so little real physical evidence, and with so many gaps filled only by speculation. It is especially troublesome, for instance, to learn that there are no known fossils of the creature that is said to have preceded *Eohippus* and that there is also a gap in the proposed sequence immediately after *Eohippus* and before its proposed descendant *Miohippus*. We are entitled to ask: What exactly is it that connects them scientifically?

The problems that have bedeviled horse paleontology also beset every other branch of the science. Indeed, the gaps in the fossil record are reflected in the living world where many major animal and plant groups are high and dry with no discernible predecessors. The development of the entire order of mammals is missing from the fossil record, from its supposed shrewlike ancestor of the late Cretaceous until modern times.

Paleontologists have produced one spectacular fossil which is said to be a clear example of a transitional form not merely from one species to another, but from reptiles to birds: the famous *Archaeopteryx* skeletons found in the limestones at Solenhofen in Bavaria. So rare and precious are the two chief specimens said to be that they are kept under guard in a bank vault, safe from harm at the hands of outraged creationists who, presumably, are thought to be plotting the theft or destruction of what they consider to be Darwinist forgeries!

Like the fossil horses, *Archaeopteryx* is an important discovery and one that appears to confirm the predictions of the Darwinist model. It seems to offer substantial evidence of a transitional form and, together with fossil horses, forms the centerpiece of most museum displays and textbook accounts. Like the horses, however, *Archaeopteryx* has formidable problems, and these have been compounded by more recent discoveries.

Darwinists believe *Archaeopteryx* is proof of a number of important parts of their theory. First, it is said to demonstrate the existence of a feathered creature long before the age of birds—*Archaeopteryx* dates from the age of reptiles. Next, it is said to have vestigial characteristics from its reptilian ancestors: claws on its feathered forelimbs, teeth in its beak, and a bony reptile-like tail. It indisputably possesses true feathers and wings, but it does not possess the large pectoral muscles and deeply keeled breastbone that would enable it to fly. It must have been either virtually flightless, like a chicken, or have been a glider—a possible precursor of true flight, say Darwinists.

Like the horse fossils, all this seems very convincing—until you subject the claims to a detailed examination. The idea that *Archaeopteryx* had descended from dinosaurs was first floated in the 1870s by Darwin's champion Thomas Huxley because of the persuasive similarities of the legs and hips of birds with those of dinosaurs. However, in offering *Archaeopteryx* as a descendant of dinosaurs, Huxley was ignoring one important inconvenient fact—*Archaeopteryx*, like all birds, has a wishbone (analogous to the mammalian clavicle or collarbone) whereas dinosaurs did not possess collarbones.

In 1926 Darwinist paleontologist Gerhard Heilman published a very detailed review of all the evidence for bird origins and carefully analyzed all the relevant anatomical questions. Heilman concluded that the most likely candidates for the ancestor of *Archaeopteryx* were dinosaurs and that, among these, the "coelosaurs" (small bipedal carnivores) were the best candidates. Unfortunately, Heilman wrote, dinosaurs did not have collarbones so a coelosaur ancestry was out of the question. Heilman proposed therefore that *Archaeopteryx* must have descended from a hypothetical pre-dinosaur ancestor from the Triassic period, and later developed fused collarbones by "convergent" evolution. This conclusion—despite its wholly hypothetical foundations and somewhat circuitous logic—passed into Darwinist lore and has been repeated in textbooks and museum displays ever since.

Matters rested there until 1973, when Professor John Ostrom of Princeton resurrected the idea that birds have descended from coelosaurs. Ostrom made a detailed anatomical analysis of *Archaeopteryx* and found some twenty points of similarity with coelosaurs. Moreover, further collecting had shown that a few dinosaurs did exist with collarbones, so perhaps some coelosaurs or close relatives might have had collarbones too. However, according to Dr. David Norman in his *Illustrated History of Dinosaurs:*

> Dr. Sam Tarsitano and Dr. Max Hecht are recent advocates of Heilman's original proposals of a more distant Triassic archosaur ancestor of birds. They claim to have found major faults with Ostrom's original work. Also several embryologists claim that the three fingers of the modified hand of living birds could not possibly have evolved from the three fingers of the theropod hand because the hand of birds is composed of the 2nd, 3rd and 4th fingers while in theropods the fingers are the 1st, 2nd and 3rd! Quite where this leaves *Archaeopteryx*, which also appears to have a theropod-like hand of fingers 1, 2 and 3, is a matter of some embarrassment—does it mean that *Archaeopteryx* was merely a feathered dinosaur and not related to birds at all?[5]

Archaeopteryx has recently been subjected to even more embarrassment since it has lost its title as the earliest bird (if bird it is). Sankar Chatterjee, professor of paleontology at Texas Tech University,

Archaeopteryx fossil (left) and restoration (below). *Archaeopteryx* is said by Darwinists to have evolved from dinosaurs called "coelosaurs" and to be ancestral to modern birds. But coelosaurs did not have collar bones while *Archaeopteryx* did. And while birds' wings are composed of the second, third, and fourth fingers of the hand, *Archaeopteryx's* wing is composed of the first, second, and third fingers. *(Photos: Natural History Museum, London)*

described a newly discovered fossil bird in the July 1991 Philosophical Transactions of the Royal Society. The new fossils, called *Protoavis texensis*, are those of a creature the size of a pheasant which was undoubtedly capable of flapping flight. They come from beds in Texas said to be 75 million years older than those in which *Archaeopteryx* was found.[6]

This means that true birds, essentially the same as modern birds, were flying happily in the skies of Texas during the period that Darwinists like to call the age of reptiles, a further indication that their geochronometry may well be faulty and that birds and extinct reptiles were in fact contemporary in a more recent past.

There are other problems with *Archaeopteryx*, too. The possession of claws on its wings is not diagnostic of reptilian ancestry, nor is it unique to *Archaeopteryx* since there is a modern bird in Venezuela, the hoatzin, which when young has such claws on its winged forelimbs. The wing claws of both *Archaeopteryx* and the hoatzin are sometimes referred to by Darwinists as "vestigial," but no evidence as to what creatures they are descended from and hence what precisely the claws are vestiges of has been produced. Teeth and a bony reptile-like tail certainly are unique characteristics for a bird, but it is no longer certain that *Archaeopteryx* is a bird.

So although *Archaeopteryx* is undoubtedly a fossil discovery of some significance, it is quite impossible to say at present exactly what that significance is. More importantly, it is impossible for Darwinists to claim that it supports the mechanism of random genetic mutation coupled with natural selection. *Archaeopteryx* provides no evidence for either mechanism, since it is completely isolated in the fossil record, just like *Eohippus*, with no known direct predecessor and no known direct descendant.

Darwinists have dealt with the lack of real transitions in the fossil record in two ways, both of which seem perfectly reasonable. First they have said that all vertebrate fossil remains are relatively rare and finds depend largely on chance. The fact that a particular specimen has not yet been discovered does not rule out the possibility of its being found at some future date.

Darwin himself raised this point in connection with the lack of fossil remains of early humans, showing part-ape and part-human characteristics, when he observed in *The Descent of Man* that

With respect to the absence of fossil remains serving to connect man with his ape-like progenitors, no-one will lay much stress on this fact who reads Sir C. Lyell's discussion, where he shows that in all the vertebrate classes the discovery of fossil remains has been a very slow and fortuitous process. Nor should it be forgotten that those regions which are the most likely to afford remains connecting man with some extinct ape-like creature, have not as yet been searched by geologists.

In fact, more than 100 years of further intense collecting by well-funded professional expeditions has not yet yielded any of the remains that Darwin envisaged, and Africa and the Middle East (the areas "most likely") have now been thoroughly searched. There are early apelike remains and there are early hominid remains. Indeed the store of primate fossils has been multiplied a thousand-fold since Darwin. But the only "missing link" so far discovered is the bogus Piltdown man, where a practical joker associated the jaw of an orangutan with the skull of a human.

Darwin also gloomily confessed in *The Origin of Species* that

> The number of intermediate varieties which have formerly existed on Earth must be truly enormous. Why then is not every geological formation and every stratum full of such intermediate links? Geology assuredly does not reveal any such finely graduated organic chain; and this, perhaps, is the most obvious and gravest objection which can be urged against my theory.

Some evolutionists have explained the absence of transitional remains by suggesting that evolution proceeds in fits and starts. A species like *Eohippus* could remain stable for a long time—perhaps millions of years—thus giving rise to many individuals, some of whose bodies are fossilized. But then there is a spurt of evolutionary activity and *Eohippus* relatively quickly mutates into *Mesohippus* which again remains stable for millions of years and gives rise to many fossil remains.

Again, any reasonable person can hold this view quite properly.

But as before, he cannot hold it by virtue of the evidence of the geological record, because there is no paleontological evidence for such evolution in bursts—except the lack of transitional fossils which was the very reason for the existence of this point of view.

An exasperated Darwinist may well feel entitled to ask: If you won't accept the *Eohippus* sequence or *Archaeopteryx* as evidence for transitions, what on earth will you accept?

The answer could not be easier. Three-quarters of the Earth's land surface is covered with sedimentary rocks. A great proportion of these rocks are continuously stratified where they outcrop and the strata contain distinctive fossils such as sea urchins in the chalk and ammonites in many Mesozoic rocks. The case for Darwinism would be made convincingly if someone were to produce a sequence of fossils from a sequence of adjacent strata (such as ammonite species or sea urchins) showing indisputable signs of gradual progressive change on the same basic stock, but above the species level (as opposed to subspecific variation). Ideally this should be demonstrated in a long sequence, ten or twenty or fifty successive fossil species, showing major generic evolution—but a short sequence would be enough.

But this simple relationship is not what is shown in the sequence of the rocks. Nowhere in the world has anyone met this simple evidential criterion with a straightforward fossil sequence from successive strata. Yet there are so many billions of fossils available from so many thousands of strata that the failure to meet this modest demand is inexplicable if evolution has taken place in the way Darwin and his followers have envisaged. It ought to be relatively easy to assemble not merely a handful but hundreds of species arranged in lineal descent. Schoolchildren should be able to do this on an afternoon's nature study trip to the local quarry, but even the world's foremost paleontologists have failed to do so with the whole Earth to choose from and the resources of the world's greatest universities at their disposal.

A few miles south of London a stiff blue clay called the gault is quarried. It has been dug by brick makers for hundreds of years and, once fired, gives London's Georgian houses their distinctive yellow brick. This useful deposit is historically important also.

The authors of the *Geological Survey of Great Britain* memoir on Folkestone say,

Nowhere is the gault more readily accessible, its fossils more abundant or more perfectly preserved than in the cliffs and shore of Folkestone. The descriptions of the gault cliffs at Copt Point by De Rance and Price constituted one of the earliest uses of paleontology for stratal subdivision on modern lines. Added to the more recent researches of L. F. Spath, these investigations have raised the gault succession at Folkestone to the status of an international yard-stick for middle and upper Albian times.[7]

The authors then reproduce a detailed table of ammonite zones compiled by Spath (1923 and 1942) and modified by Breistroffer (1947) and Casey (1949). This table lists fourteen successive beds distinguished by ammonites in four major zones. The four zones are called after the ammonites they contain: the lowest zone is called the *dentatus* zone, the next the *lautus* zone, the third the *inflatum* zone, and last the *dispar* zone.

The species of ammonite associated with these zones can be collected by the thousand. Museums and private collections are full of them, preserved in beautiful detail including an iridescent pearly shell. They come from a section of clay perhaps 100 feet high, which presumably, in uniformitarian terms, represents millions of years of sedimentation. Yet among the tens of thousands of specimens dug up by collectors, no one has ever found a specimen that is part way between *Hoplites dentatus* and *Euhoplites lautus* or between *lautus* and *Mortoniceras inflatum*—or between any of the fourteen different ammonites.

There are plenty of other ammonites in the clay, of all shapes and sizes—which Darwinist geologists often describe as separate species. Unfortunately, however, these do not fall neatly in between the primary species in their anatomy, nor neatly in the clay strata between them. They do not show evolution in a straight line but, like the horses, "fall into the usual phylogenetic net." Ammonites get more ribs then less ribs; they become closely coiled then loosely coiled; they grow lumps, become smooth, then grow lumps again.

This one example can be multiplied by all the quarries and all the sea cliffs, all the road cuttings and canal ways and railway embankments in the world. Wherever there are successive strata

Fossil ammonites, from the gault clay of Kent. Top: *Douvilleiceras, Beudanticeras, Hoplites*. Bottom: *Euhoplites, Anahoplites, Dimorphoplites*, and *Mortoniceras*. Darwinists believe they are an evolutionary sequence but there are no intermediate forms in the beds between. *(Photo: author's collection)*

containing distinctive species, no one has ever demonstrated an unmistakable line of descent. Indeed, one thing that becomes plain to the open-minded geologist as he travels from exposure to exposure in search of fossils is that nature almost perversely precedes and follows one species by quite different ones.

Some Darwinists have even attempted to press this perversity into serving as evidence for their theory. For example, in the Cotswold hills, near Gloucester, there is a large brickpit in the village of Blockley. The bluish clay at Blockley looks like the gault of Folkestone but is actually an earlier formation known as the Lias, dating from the Jurassic period. The Liassic clays at Blockley provide many well-preserved ammonites, which are used as zone index fossils.

There are two main kinds of ammonite found at Blockley. There are fat ones with two rows of knobs on the side (called *Liparoceras*) and thinner ones with no knobs (called *Aegoceras*). Occasionally, collectors have also found a third kind which is said to be intermediate between these two and which has been called *Androgynoceras*. This third kind resembles *Aegoceras* in its inner whorls (that is, when it was young and the shell was first forming) but later on resembles *Liparoceras* with its two rows of knobs.

In 1870 a dedicated Darwinist named Hyatt took the specimens in the British Museum and arranged them in the order *Aegoceras* (oldest)–*Androgynoceras*–*Liparoceras* (youngest). His reasoning was based on the then fashionable evolutionary theory of "recapitulation," the supposed repetition of characteristics in later generations. In 1938 the distinguished paleontologist L. F. Spath took the same specimens from the British Museum and after careful examination reversed the order of evolution: *Liparoceras*–*Androgynoceras*–*Aegoceras*. Careful collecting in Dorset had revealed that the oldest (lowest) beds yielded only *Liparoceras*, while *Aegoceras* was found only in the youngest beds, accompanied by occasional *Liparoceras*. In reaching this conclusion, Spath was employing another evolutionary concept that was very different from that used by Hyatt. This time it was called "proterogenesis" and was said to concern the appearance of new characteristics only in the young, which then later spread to the outer parts of the shell.

In 1963 the ammonites from Blockley, where all three types are found in sequence, were examined again by Callomon. On this occasion neither of the previous explanations was found to be satisfactory and the variation of *Androgynoceras* was attributed instead to "extreme sexual dimorphism"—meaning that male and female of the same species were of radically different shape.[8]

Naturally, one sympathizes with the difficulties of the paleontologist attempting to identify the sex of a creature whose last night on the town (according to uniformitarians) was some 150 million years ago. No one can blame a researcher who makes a mistake that is rectified by further research, for this is the very method of science. What I believe the Blockley ammonites demonstrate is something else. It is a prime illustration of the infinite elasticity of Darwinian theory: of its ability to interpret the data in any one of a number of completely different ways—even with diametrically opposed conclusions—as long as those ways are consistent with the central belief in Darwinian evolution itself.

"Recapitulation" means the fossils evolved one way. "Proterogenesis" means they evolved in the opposite direction. In reality

Fossil ammonites from the Lias of Blockley. *Liparoceras* (left) and *Androgynoceras* (right). Darwinists have variously claimed that the species on the left is the ancestor of the one on the right; that the species on the right is the ancestor of the one on the left; and that the two forms are male and female of the same species. Darwinist theory can accommodate all three conclusions. *(Photo: author's collection)*

the Blockley ammonites give no clue to lineage at all, just like all the other ammonites from all the other quarries.

Probably the most ambitious and comprehensive work on paleontology ever to be published is the series of volumes produced in the 1950s by the Geological Society of America and the University of Kansas Press under the guidance of a committee of the most distinguished paleontologists in the English-speaking world. Under the title *Treatise on Invertebrate Paleontology* some twenty-four volumes draw together the sum of human knowledge on thousands of fossil species. If solid fossil evidence of evolution is to be found at all, it is to be found documented in the Treatise's volume that deals with the richest of all fossil fauna, the ammonites. In Volume L are illustrated and described in minute detail hundreds of ammonite species. Yet under the promising heading of "Examples of Ammonoid Evolution" the editor issues this warning to readers keen to learn what proof the fossil record has to offer:

> Waagen (1869) in a pioneer work attempted to demonstrate lineages, or lines of descent. . . . The chief obstacle to such studies is that a lineage is an oversimplified concept; it is impossible to pick out a stratified succession of individuals which can with certainty be said to be genetically connected in the strict ancestor-descendant relationship.[9]

Having warned readers that the process is "impossible," the editor then moves on to quote the works of Spath and Howarth as making just such a connection. As is always the case, the descriptions, the evolutionary reasoning offered, and the suggested lines of descent are little more than subjective value judgements.

Of course, this is not to say that the findings of comparative anatomy are without foundation—quite the contrary. It was the very fact that animals as diverse as the mouse and the elephant both display a similar four-limbed anatomical pattern that first led biologists to think that these animals might have a common ancestor.

These and many other examples of similarities in form seem to point unequivocally to common ancestry and common processes of evolutionary change. The problem with this view is that the apparent kinship relationships are, at least to some extent, an artifact of the system employed by science to describe and classify each species.

The system of taxonomy or zoological classification which provides us with the concepts of "species," "genus," "family" and with the classification of animals into "orders" such as mammals and reptiles, is something that we take very much for granted today, to the extent that it has become absorbed in our everyday language. The system was devised some two hundred years ago by the Swedish naturalist Carl Linné, and has subsequently been adopted by the international scientific community. The modern Linnaean system of classification is bound by very strict rules governing the admission of each newly discovered animal and plant to the catalog of species. Immense care is given to the smallest detail of nomenclature and its application, so that the system will not fall into disrepute through misuse. Yet despite the enormous utilitarian value of the system in providing a common international language for naturalists, and its great usefulness in cataloging the plant and animal kingdoms, the system cannot legitimately be used—as Darwinists often wish—to decide the question of the kinship, if any, between those plants and animals.

The question of whether biological types are real or exist only as labels, a mere by-product of human observation, is an ancient debate, nominalists versus realists, that stretches back to Plato's time. As far as biology and evolution theory are concerned, the debate remains unsettled because, as Norman Macbeth has pointed out, nature herself capriciously provides evidence for and against both sides. Those who believe species are real can point to examples like the ginkgo tree which stands in magnificent isolation with no relatives and an unchanging form throughout the geological record. Those who believe species are merely convenient labels point to the willow trees, of which there are countless varieties which blend into each other and are impossible to differentiate.[10]

In biology the debate between realists and nominalists has not been settled but has degenerated into a kind of uneasy truce in which the philosophical issue has been quietly forgotten and replaced by a purely empirical approach.

One of the twentieth century's greatest authorities on taxonomy is Ernst Mayr, Harvard's professor of zoology. Mayr's standard work, *Principles of Systematic Zoology*, admits that all categories such as "genus" and "family" are quite arbitrary in that they seek to describe relationships which cannot be demonstrated experimentally with living populations. Nature is so complex, says Mayr, and so inconsistent that "no system of nomenclature and no hierarchy of systematic categories is able to represent adequately the complicated set of interrelationships and divergences found in nature."[11]

Mayr and his fellow leaders of the synthetic school, Dobzhansky and Simpson, solved the problem by rejecting any attempt to elaborate a theory of biological types, and substituting instead a theory of breeding populations. A population consists of a single species when its members interbreed—producing fertile young—with each other but not with other such breeding populations.

Whatever the outcome of the philosophical debate between nominalists and realists, biologists of all persuasions have rejected taxonomy except as a mere convenience when referring to animals and plants. And once the taxonomic categories of "species," "genus," "family," and the like are admitted to be no more than convenient metaphors or contrivances, then we are left simply with a biological realm that consists of individuals—all of which are different.

For example, you might imagine that all human beings are constituted in exactly the same way the world over. But, surprisingly, this is not the case. Humans vary considerably in matters of anatomical detail such as the number of fingers and toes; structure of internal organs like the stomach; the number of bones in the wrist; the number of pairs of ribs (eleven, twelve, or thirteen) and other features like the amount of body hair and webbed skin between fingers or toes. Usually we marginalize these variations by clinging to a concept of a "normal" anatomy and dismissing differences as being freaks of nature. In fact, from a genetic standpoint, every organism is unique.

What this means is that the tables drawn up by biologists to classify animals and plants are quite different in character from the tables drawn up by physicists to classify the chemical elements, for example. In the case of the periodic table, each element has uniquely identifiable physical properties (the number of particles contained in the atom of that element) and behaves in a predictable way whenever or wherever an experiment is conducted with that element. In the case of the Linnaean system of classification, the majority of living things, and fossils, are distinguished mainly in a statistical rather than absolute way; and sometimes their behavior is experimentally predictable, sometimes it is not.

This is the reason that comparative anatomy must be treated with the greatest care when it is used to deduce "evolutionary" relationships, which actually rely on the taxonomic system rather than blood. The tooth that Gideon Mantell eventually identified by "coincidence" did not come from a true relative of the iguana, but nevertheless, the animal was called "iguanodon" and the characteristics of the iguana were wished upon the extinct animal: It must have been a reptile; hence it must have been cold-blooded; hence it must have been sluggish, and so on—characteristics which are doubted by some specialists today, such as Robert Bakker of the University of Colorado and Nicholas Hotton of the Smithsonian Institute, both of whom argue that dinosaurs like iguanodon were warm-blooded.[12]

The confusion into which some Darwinists have led science by bending the system of zoological classification to fit their theory was pointed out by W. R. Thompson, director of the Commonwealth Institute for Biological Control in Ottawa, who wrote the introduction to a centenary edition of Darwin's *The Origin of Species*:

> The general tendency to eliminate, by means of unverifiable speculations, the limits of the categories nature presents to us, is the inheritance of biology from *The Origin of Species*. To establish the continuity required by theory, historical arguments are invoked, even though historical evidence is lacking. Thus are engendered those fragile towers of hypotheses based on hypotheses, where fact and fiction mingle in an inextricable confusion.[13]

If taxonomy and its handmaiden, comparative anatomy, are misleading in the living world, they are doubly dangerous when applied to fossils. When the zoologist studies the anatomy of living creatures and compares them, he has available evidence not only of the hard parts such as bones and teeth, but also of the structure of internal organs, the composition of blood, evidence of skin, hair, coloration, and of processes and functions such as body temperature and method of reproduction, none of which survive in fossil form (except in a few freak cases). Often these nonsurviving features—whether it is cold-blooded or warm-blooded, whether it bears live young or lays eggs, and what kind of food it lives on—are crucial in describing an animal. As noted above, using taxonomy, evolutionists ascribed reptilian characteristics such as cold-bloodedness to dinosaurs, a belief now doubted by some paleontologists.

Whereas the zoologist can base his description of animals and their living habits on the whole range of its characteristics, the paleontologist has only the hard parts, bones and teeth, on which to form a judgement. This need not be an insuperable obstacle to accurate diagnosis. Indeed in the field of forensic medicine we are accustomed to police pathologists performing seeming miracles of detective work in identifying the sex, age, and height of dismembered skeletons, and eventually even establishing the identity and cause of death of the victim. In a few cases such evidence has even led to conviction of the murderer. Professor Keith Simpson solved a case of this kind as a young pathologist during the Second World War. In 1942 workmen demolishing an old Baptist church in South London uncovered a skeleton from which the arms had been cut off below the elbow and the legs removed below the knee. Simpson identified the skeleton as that of a woman by the size of the hip joints and used Pearson's formula and Rollet's tables to estimate the height as 5 feet ½ inch. X-ray photographs of the skull plates showed that the brow plates were completely fused while the top plates were in the process of fusion, which put her age at between 40 and 50 years. Teeth in the upper jaw had been filled and this enabled the dead woman to be identified by dental records. Simpson's detective work placed the dead woman's husband in the dock, accused him of her murder, and he was convicted.

Similar feats of identification have been performed on human skeletons from remote historical times, such as that of Cleopatra, the wife of Philip II, King of Macedonia, found in 1987. Examples can also be found in paleontology, where remarkable anatomical descriptions can be given from what appears to be the most meager information, especially teeth.

There is, though, a crucial difference between scientific detective work which provides the basis for a case and scientific evidence which is actually used to make a case. Keith Simpson's investigation provided evidence of identification and pointed the finger at murder. From this beginning the police were able to build a case against the husband by uncovering evidence of motive, means, and opportunity. Similarly, the Greek archeologists excavating in the town of Vergina (capital of ancient Macedonia) were able to provide evidence only of the interment of a young woman in her early twenties. It is the circumstances of her burial, such as great treasures and other circumstantial evidence, that point to her being Philip's wife. When a paleontologist carries out his detective work on the lithified shell of an ammonite, he is compelled to employ intelligent conjecture from the outset. He is using the *circumstances* of the burial as direct evidence of the nature of the extinct creature that used to inhabit the shell. For example, the chalk in which Cretaceous ammonites are found is believed to have been deposited in a warm, shallow sea. Thus the creatures found in the chalk will be expected to have been of the type who thrive in such shallow conditions. Obviously, the kind of circumstantial evidence and the kind of conjecture employed by the paleontologist will depend on the assumptions he has already made about how the rocks were deposited and what pattern of life he expects to see in those rocks.

This kind of detective work is almost the opposite of that employed by the forensic scientist. Instead of providing a firm foundation of scientific fact on which the biologist may conjecture the various surrounding circumstances, the paleontologist, who is a convinced Darwinian, is providing a basis of conjecture on which the biologist may erect further conjectures—the "fragile towers of hypothesis on hypothesis" referred to above by Thompson. If comparative anatomy is unhelpful (or misleading) in seeking to substantiate synthetic evolution, let us turn to one field where Darwinists must feel absolutely secure: the very heart of the theory, its central mechanism, natural selection.

PART THREE

CHANCE

CHAPTER 11

Survival of the Fittest

W HEN HE BORROWED THE PHRASE "survival of the fittest" from Herbert Spencer, Darwin made it clear that he intended it to mean precisely the same thing as his own memorable phrase "natural selection." "This preservation of favorable individual differences and variations, and the destruction of those which are injurious, I have called Natural Selection, or the Survival of the Fittest," he wrote in *The Origin of Species*.

The concept of natural selection is fundamental to Darwinian evolution theory. Coupled with random mutation, it is the one and only mechanism proposed to account for changes in form fitting a species, sometimes uniquely, to its mode of life—the streamlining of the dolphin or the giraffe's long neck. According to Julian Huxley, "So far as we now know, not only is natural selection inevitable, not only is it *an* effective agency of evolution, but it is *the* only effective agency of evolution."

The giraffe has a long neck, according to Darwinists, for three reasons. First, because an ancestral animal experienced a mutation which fortuitously gave it a longer neck; second, because the longer neck gave it some competitive advantage (such as being able to feed higher up the tree) so it survived to produce many offspring; and, third, because this natural advantage also favored its descendants, a majority of which would inherit the long neck. The second two stages of this process are what Darwin meant by his phrase.

Darwin also saw natural selection taking place in a hostile environment where the majority of offspring die before reaching maturity or breeding. This view, the core of Darwinian thinking, was summed up by modern synthetic evolutionists such as Huxley, Mayr, and Simpson in the phrase "differential reproduction" as being synonymous with natural selection.

As natural selection or differential reproduction is such an important mechanism, you might expect to find a large body of technical literature on the subject, with many detailed studies and observations from the natural world. Regrettably, you will search the world's scientific libraries in vain for such studies because it turns out—for reasons examined in detail a little later—that natural selection cannot be studied in any experimental way.

Natural selection means those animals and plants that are best fitted to their environment and way of life are the most successful. How do we measure or evaluate the fitness of an animal or plant? By its capacity to survive, say Darwinists. How is "survival" measured? By the number of offspring left. So, fitness means breeding success. But survival is also measured by breeding success. Restated, the "survival of the fittest" means: the prolific breeding of the most prolific breeders. Put this way, does natural selection mean anything at all?

Waddington answered this question in 1960 when he wrote:

> Darwin's major contribution was, of course, the suggestion that evolution can be explained by the natural selection of random variations. Natural selection, which was at first considered as though it were a hypothesis that was in need of experimental or observational confirmation, turns out on closer inspection to be a tautology, a statement of an inevitable although previously unrecognised relation. It states that the fittest individuals in a population (defined as those which leave most offspring) will leave most offspring. Once the statement is made, its truth is apparent. This fact in no way reduces the magnitude of Darwin's achievement; only after it was clearly formulated, could biologists realise the enormous power of the principle as a weapon of explanation.[1]

Many will be surprised to find a professor of biology describing a tautology as an achievement of any sort. Waddington failed to recognize the damaging nature of his admission that "natural selection" and its synonymous phrase "survival of the fittest" are nothing more than tautologies. However, he more than made up for it by his prescience in accurately foreseeing how Darwinists would use natural selection—as a "powerful weapon of explanation."

Darwin conceived his idea of natural selection by analogy with *artificial* selection, something which—as a capable animal breeder himself—he knew a great deal about. He bred pigeons and other animals and traveled extensively in Southern England discussing animal husbandry with other breeders. He knew that it was possible for the stockbreeder to change the characteristics of an animal—the dairy cow or the sheep, for example—quite substantially in only a few generations.

If humans can change an animal's characteristics by selection, in only a few years, Darwin wondered, what could nature not achieve in millions of years? Instead of the hand of the stockbreeder, using his experience and his judgment to pick the characteristics he wanted, nature herself, acting through the harsh realities of the competitive environment, was selecting precisely those characteristics that conferred advantages for continued existence and propagation of offspring, thus ensuring "the survival of the fittest."

On first acquaintance, this starkly noble idea appears irreducibly simple. On closer inspection, it is found to be a densely compressed complex of tacit assumptions, few of which correspond with observations of the natural world.

Talk of "survival" immediately conjures up a lurid vision of competition between the various forms of animal life in a hostile world: competition for the scarce resources of food and living space, of "nature red in tooth and claw" as Tennyson pictured it for his enthralled Victorian readers. In reality such competition is very rarely found in nature. One conservative estimate is that there are at least 22,000 common species of fish, amphibians, reptiles, mammals, and birds. In addition there are said to be at least 1 million common insect species. Some of these thousands of species—notably humans—do compete aggressively, killing competitors for living space and food. But the species that do are very much in the minority. The overwhelming majority of creatures do not fight, do not kill for food and do not compete aggressively for space in a way that results in the "loser" dying out.

Earlier in the century it was widely accepted that this kind of behavior took place with measurable effects on survival, largely because the evidence was misinterpreted. The male fiddler crab, for instance, has one enormous claw and one normal-sized claw which it uses to eat. It was assumed that the enormous claw was to fight its fellow males for the privilege of mating with the most desirable females and possessing the most desirable territory. Observation of male fiddler crabs, however, shows that they do not use their large claw to fight. Indeed, they seem to signal the presence of food to their fellow crabs. Far from being a weapon of war, the fearsome claw is an instrument of social cooperation.

There are scores of similar cases where attributes or behavior were assumed to be aggressive but detailed observation has shown these assumptions are unfounded. Fighting between males for "domination" generally leads to no particular advantage for the "winner." His opponent simply goes elsewhere and mates. Often also, the females will mate as readily with the loser as with the winner. The fighting rarely results in any fatal injury, and seems to be much more ritual than actual (rather like the fighting of teenage boys in fact). Often when fatal injuries occur in intraspecies conflict, it is the result of accidents, such as when the antlers of male deer become locked, and is fatal to both parties.

The origin of this idea of the struggle for existence resulting in a culling of the less well adapted was Thomas Malthus's *Essay on Population* published in 1798. Darwin was deeply impressed by Malthus's conclusion that nature automatically regulates the size of human populations through the food supply. He made this Malthusian mechanism the starting point of his theory:

A struggle for existence inevitably follows from the high rate at which organic beings tend to increase. . . . Hence as more individuals are produced than can possibly survive, there must in every case be a struggle for existence, either one individual with another of the same species, or with the individuals of distinct species, or with the physical conditions of life. It is the doctrine of Malthus applied with manifold force to the whole animal and vegetable king- doms; for in this case there can be no artificial increase of food and no prudential restraint from marriage.

Darwin goes on to give many examples of the nature of "checks to increase," including the destruction of seedlings by a variety of en- emies and the effect of climate on birds' nests (he estimated that the winter of 1854-55 destroyed 80 percent of the birds on the grounds of his home at Down House in Kent). He also stressed, however, that the exact causes of these checks are often obscure.

The key point about this belief from the Darwinist point of view is that it is another example of nature acting blindly. There is nothing, says Darwin, that the animal or plant populations can do to affect the consequences of overpopulation—"no artificial increase of food and no prudential restraint from marriage." Death of those not well fitted to exist is the inevitable result.

Today natural history museums make this matter the principal plank of their exhibits on Darwinism. Display cases show models of rapidly breeding animals such as rabbits and explain that the need for rabbits to have a territory, grass to forage, ground for burrows, and the need to keep clear of predators limits the avail- able space and hence keeps the rabbit population in check. Similar arguments are applied to other animals and to plants as well. These ideas do indeed contain obvious truisms. The question is, do they contain the great principle that Darwinists believe?

It is certainly true that many creatures are at risk of losing their lives to hardship and to predators—sometimes even predators of their own species. But this form of conflict does not necessarily lead to the kind of competition that would promote a favored few. The majority of carnivores do not feed on prey that they themselves have just killed but rather are scavengers or carrion feeders. This includes legendary hunters such as lions or sharks who frequently eat as a result of not their own direct efforts but those of another lion or shark. (This is also true of humans.) Thus the "successful survivor" is not necessar- ily the most capable hunter-killer and does not necessarily possess the characteristics of such a killer. It follows that if these characteris- tics are not present, they will not be preserved by breeding.

The Darwinian concept contained another important tacit assumption: that it is within the power of individuals to take action to ensure their survival. That, for instance, the toughest, cleverest, most determined, and most enterprising lion will ensure its survival by seeking out new territory and new sources of food when the prey runs out in its usual territory. But, of course, in many cases it is likely that prey has run short because of some natural calamity such as drought, fire, or flood. Even if the lion escapes the immediate disaster, there simply may not be alternative sources of food and no action it takes can affect its survival. By the same token, an unenterprising, cowardly, stupid predator in another part of the world may escape the drought or other natural calamity, survive, and breed. So it is not the "fittest" that survives but the luckiest—a quality which is not usually thought of as inheritable.

But it is not merely observation of details that is faulty in the concept of the survival of the fittest, it is the concept itself. Why should aggressive competition, in which the vanquished fail and die and the victorious survive and prosper, be beneficial for the race? As described earlier a whole host of factors, including chance, plays a part in the success of an individual of any species. Whether a seed falls on fertile or stony ground is a matter of luck. There is no mutation that can assist a sycamore seed to germinate and grow on a wave-washed bare rock.

The concept that the harsh action of the competitive environment is a valuable process that strengthens the breed and weeds out weaklings was a tacit part of the nineteenth-century view of evolution. Nature was a gigantic health club, forcing each species to shape up or ship out. If, in this harsh process, the weak went to the wall then that was too bad. It is merely nature's way of ensuring that only the fit survive.

The source of this view is nature's indifferent cruelty to the millions of individuals of some species who are born but who perish before attaining maturity and mating. There are only a limited number of winning tickets. Those who fail to grasp such a ticket are doomed to die. Only the toughest and cleverest can wrest a passport to life from nature's cruel grip.

The concept of the struggle for existence, though central to evolution theory in the nineteenth century, has receded in importance;

today it is rejected as being either a noncontributory factor in evolution or actually detrimental to it. Simpson wrote:

> Struggle is sometimes involved, but it usually is not, and when it is, it may even work against rather than toward natural selection. Advantage in differential reproduction is usually a peaceful process in which the concept of struggle is really irrelevant. It more often involves such things as better integration into the ecological situation, maintenance of a balance of nature, more efficient utilization of available food, better care of the young, elimination of intra-group discord (struggles) that might hamper reproduction, exploitation of environmental possibilities that are not the objects of competition or are less effectively exploited by others."[2]

For Julian Huxley life's struggle is no more than a banal observation of little significance. "The struggle for existence," Huxley wrote, "merely signifies that a portion of each generation is bound to die before it can reproduce itself."[3]

The modern position therefore is that natural selection and the survival of the fittest are no more than empty tautologies, while the struggle for survival plays no important part in evolution. This loss of any real significance in Darwin's central concept left synthetic evolutionists in a hole from a theoretical standpoint. Harvard's George Simpson attempted to restore some scientific content to the concept thus:

> If genetically red-haired parents have, on average, a larger proportion of children than blondes or brunettes, then evolution will be in the direction of red hair. If genetically left-handed parents have more children, evolution will be towards left-handedness. The characteristics themselves do not directly matter at all. All that matters is who leaves more descendants over the generations. Natural selection favors fitness only if you define fitness as leaving more descendants. In fact geneticists do define it that way, which may be confusing to others. To a geneticist, fitness has nothing

to do with health, strength, good looks or anything but effectiveness in breeding.[4]

This sounds solid enough, and it certainly avoids all the old pitfalls. Natural selection is the process by which the most successful breeders populate the world, and the less successful breeders die out—*regardless* of their respective characteristics. Let us go back to first principles and apply this formula. The giraffe has a long neck because . . . ? Here we get stuck. The only help we get from synthetic evolution is that the giraffe has survived because it has survived. Natural selection is unable to offer any evidence or insight into its evolution because "the characteristics themselves do not directly matter at all."

What this really means is that Darwinists have become reluctant to try to explain any particular characteristic as being responsible for the giraffe's evolution—even regarding its long neck—because they would then have to show *how* and *why* that characteristic has favored the giraffe over other animals, some of whom are extinct. Natural selection has proved a completely inadequate tool for such explanation since it does not allow us to refer to individual characteristics at all. All that Darwinists dare say with impunity is that the giraffe has survived because it is "adapted" to its environment—the modern way of expressing the old tautology.

To summarize, the modern position of the synthetic theory is: the struggle for existence plays no part in evolution. The direction of evolution is determined solely by the characteristics of those animals and plants that are successful breeders. We are unable to say anything about why a particular characteristic might favor, or prejudice, the survival of any particular animal or plant.

Thus "the survival of the fittest," or "natural selection," or "differential reproduction" sheds no light on the mechanism of evolution and is only another way of saying that some animals survive and prosper while others die out—an observation of limited value.

Perhaps an even more damaging criticism of the concept of natural selection is that—limited though its content may be—it is so nebulous that it can be made to fit a whole range of mutually contradictory outcomes of the evolutionary process. Natural selection is entirely compatible with the notion that all organisms in stable environments have reached a fitness peak on which they will remain forever. At the same time natural selection is entirely compatible with the idea that all organisms should regress to the safest common denominator, a single-celled organism, and thus become optimally adapted to every habitat.

In precisely the same way, because of its infinitely elastic definition, natural selection can be made to explain opposed and even mutually contradictory individual adaptations. For example, Darwinists claim that camouflage coloring and mimicry (as in leaf insects) is adaptive and will be selected for, yet they also claim that warning coloration (the wasp's stripes) is adaptive and will be selected for. Yet if both propositions are true, *any* kind of coloration will have *some* adaptive value, whether it is partly camouflage or partly warning, and will be selected for.

As a theory, natural selection makes no unique predictions but instead is used retrospectively to explain every outcome: and a theory that explains everything in this way, explains nothing. Natural selection is not a mechanism: it is a rationalization after the fact.

It could be argued that it is unreasonable to expect Darwinists to have answers to every question when so many issues in biology remain unsettled. Perhaps instead we should ask them to show us a concrete practical example of natural selection—is there such an example available? Indeed, there is. Every modern textbook with a chapter on selection and evolution and every modern encyclopedia contains extended reference to just such an example—the subject of industrial melanism in moths.

The story, as it is usually told, can be summarized as follows. In the first half of the nineteenth century the bark of trees in the Manchester area was progressively darkened by atmospheric pollution from factory chimneys. This gradual darkening affected the peppered moth, *Biston betularia*, which is nocturnal but spends the day resting with wings outspread on the trunks of trees. Before pollution the moth was light gray in color with dark gray speckles, giving it a perfect camouflage against predatory birds, since the tree trunks were also light gray. As the trunks became darker and darker, the moth "evolved" a darker protective coloration until, by 1898, some 99 percent of the moths in the Greater Manchester area had the dark coloration.

This phenomenon has been dubbed industrial melanism by Darwinists and is described (for example by the British Museum of Natural History in 1970) as "the most striking evolutionary change actually witnessed" and as "demonstrating natural selection."[5] This story, if true, would be interesting evidence in favor of evolution by natural selection. However, the story turns out to be not quite what it seems.

The change in color from light gray to dark gray is perfectly real and was recorded over the years by meticulous collecting. Also, the basic idea of changing color was thoroughly tested by Bernard Kettlewell at Oxford through experiments under controlled conditions with light and dark moths. The question is what does the change represent?

Initially, around 1848, only one or a few specimens of the dark variety were collected in the Manchester area. It was assigned the position of a subspecific variety and given the varietal name *carbonaria*. As the tree trunks became darker, the light moths lost their protective camouflage, became conspicuous, and fell easy prey to birds. At the same time, the *carbonaria* variety became better and better camouflaged and began to flourish.

Simply put, first there were a few dark moths and a lot of light ones. The light ones lost their camouflage and the dark ones gained it. All the light moths were eaten, leaving only the dark ones. Far from being an example of evolution or even of natural selection, the peppered moth is an example of a shift in population. The same thing would happen in human terms if some disease were to kill off the white race but left the black race unharmed. Similar shifts in balance continually occur among animal and plant populations, where one variety flourishes at the expense of another. But this process cannot be used to explain the central proposition of neo-Darwinism—how one species can change into a completely different species.

If "industrial melanism" is an example of natural selection then Waddington was clearly correct in his assertion that natural selection turns out to be merely a tautology.

The peppered moth thus brings us to the next most important question to be considered: Just how far *can* a species vary?

CHAPTER 12

Green Mice and Blue Genes

WHEN CHARLES DARWIN WAS COMPLETING his seminal book *On the Origin of Species,* he wanted to include a concrete example of exactly what he meant by evolution by natural selection.

Darwin had accumulated hundreds of pieces of evidence that tended to support his idea, but he knew it was inevitable that skeptics would say: Show us an example of natural selection. So in the first edition of *On the Origin of Species* Darwin gave one. He said, "I can see no difficulty in a race of bears being rendered, by natural selection, more and more aquatic in their habits, with larger and larger mouths, till a creature was produced as monstrous as a whale."

Here we have Darwin's central idea of evolution in a nutshell: bears can become whales, or whale-like, given enough time and enough natural selection. One species can turn into a completely different species by natural selection alone.

In one sense Darwin's example was very well chosen, and accords very closely with what we know of both bears and whales. Many bears are already partly aquatic. Polar bears have been sighted at sea as far as 40 miles from land. Bears are omnivorous and will happily eat fish as many people will recall who have seen films of the Alaskan brown bear fishing for salmon with its paws. But more significantly, bears will eat food of any size, including, for example, ants. This suggests that they would be equally ready to eat the myriad tiny shrimps and other marine small-fry that abound in the oceans.

At the other end of Darwin's proposed evolutionary transition, whales are believed to be mammals not all that different from bears that have returned to a marine way of life. Whales give birth to live young which they suckle, while living in family groups.

So everything seems to fit Darwin's suggestion. His example seems a well-chosen one. There is just a small gap between the aquatic, omnivorous bear and the large marine mammal, the whale, and it's not too difficult for our imaginations to fill that gap.

But despite its apparent aptness, Darwin changed his mind about this example after publication and withdrew it from the second and all later editions of his book. We don't know the exact reason why he had second thoughts and withdrew the example, but I think it is not too difficult to see why he would have done so. His book had many opponents—not just religious opponents, but scientists as well—and they would have pointed out to Darwin that his example didn't do the job he hoped it would.

In the first place, it is purely hypothetical rather than actual; it is based on conjecture not on direct evidence. What I have called an apparent "small gap" in his reasoning is in fact a vast gulf in which there are no fossils of intermediate types and no other physical evidence, so the transformation that Darwin at first saw as highly probable has not in fact happened.

With hindsight it is easy to imagine that Darwin must have felt that his suggestion was not supported by evidence, was too conjectural, and ought to be quietly dropped. But in dropping his example of bears evolving into whales by natural selection, Darwin was dropping not just a marginal example which could be easily jettisoned without penalty. In rejecting the aquatic bear, he was abandoning the central proposition of his entire theory—or at the very least was publicly displaying the kind of doubts he was privately entertaining about the process.

Since dropping the example had such drastic implications for his theory, his reasons for doing so are of considerable interest. So what kind of influences caused Darwin to drop his example about bears and whales?

Darwin was advised by many scientific friends in his writing. He corresponded with and took advice from close colleagues like Charles Lyell the geologist, Thomas Huxley the biologist, and

Joseph Hooker, director of Kew Botanical Gardens. He was also a very active researcher himself. He bred pigeons and kept careful notes on all his breeding experiments. He traveled extensively, visiting other animal and plant breeders and exchanging notes with them. He amassed considerable empirical details from hundreds or even thousands of experiments.

Darwin was well aware of the one central fact that dominated all animal and plant breeding experiments—then and now. No one has ever bred a new species artificially—and both plant and animal breeders have been trying for hundreds of years, as have scientists.

The history of human attempts to breed new species is probably thousands of years old. But here are a few relatively recent examples.

In 1811 French chemist Benjamin Delessert set up a small factory at Passy and, following the example of German chemists, made the first small quantity of crystallized sugar from sugar beets. At this time cane sugar was a strategic material denied the French because of their war with the other European powers, so Napoleon was immensely impressed by this scientific achievement. He ordered no less than forty factories to be set up in France.

However, now that France had the capability to manufacture beet sugar, it urgently needed to find, or breed, a type of beet that contained the maximum amount of raw sugar. To achieve this Bonaparte enlisted the greatest botanists in France, through the *Academie des Sciences*. A program was begun to breed selectively those sugar beet plants which gave a higher-than-average yield of sugar, a program which succeeded. At first the common varieties of sugar beet contained on average only around 4 percent sugar, but this was rapidly improved —5 percent; 10 percent; 15 percent. Then things started to go wrong. At 17 percent average yield, the sugar content of the new plants stuck, and it has stayed there to this day. And, the French discovered, repeated attempts to continue crossing high-yield varieties eventually resulted in the hybrids reverting to the low yields of their ancestral stock. These early geneticists had reached some kind of barrier, but what kind?

Synthetic evolution's most eminent experimental scientist of the twentieth century was Theodosius Dobzhansky, professor of zoology at Columbia University from 1940 to 1962 and later at the University of California until the 1970s. Dobzhansky and his coworkers carried out several long and complex series of experiments, making famous the hitherto obscure fruit fly *Drosophila*. This little fly, also called the vinegar fly, is commonly found in many parts of the world and is usually seen buzzing around rotting fruit such as apples. It is of special interest to evolutionists because it is genetically very simple. The genes of every plant and animal species—the program for its offspring coded in DNA molecules—are contained in microscopic bodies called chromosomes which are contained in every cell. A human being has twenty-three pairs of chromosomes and is genetically complex. *Drosophila* is experimentally useful because it has only four pairs of chromosomes and can breed a new generation in less than a month.

The types of experiment carried out on *Drosophila* vary in detail but are basically similar. They involve selectively breeding the fly for certain visible characteristics, such as the number of bristles growing on its body or the size of its wings.

Harvard's Ernst Mayr has described one such experiment which set out to increase the number of bristles in one group, and to decrease the number in a separate group, but with both groups starting from the same stock with an average of 36 bristles. By selecting for lower-than-normal number of bristles over thirty generations, the experimenters were able to reduce the average carried by the offspring to 25 bristles. After thirty generations, however, the line became sterile and died out. The second group was selected for higher than average number of bristles and over twenty generations the average rose from 36 to 56. Again, however, sterility became so common that the experiment was wound up.

"Obviously," says Mayr, "any drastic improvement under selection must seriously deplete the store of genetic variability." And, "The most frequent correlated response of one-sided selection is a drop in general fitness. This plagues virtually every breeding experiment."[1]

This limit to the amount of genetic variability available in a species, Mayr termed "genetic homeostasis." It is the natural barrier encountered not only by geneticists attempting to breed fruit flies, and the French botanists attempting to increase the sugar content of the beetroot, but by all plant and animal breeders throughout the ages.

Darwin himself, as a breeder of pigeons and other animals, was aware that the amount of variability available was limited. And although Darwin afterwards thought better of his statement about bears and whales, and removed it from later editions of his book, the substance of his claim nevertheless remains *the* central tenet of synthetic evolution—bears can become whales, or microbes can become elephants, by means of random mutation and natural selection. Today, few Darwinists could be found to put their names to such a bald manifesto. Yet that is what they teach in schools and universities.

Darwin's choice of example is all the more strange because he actually refers to the barrier to variation himself. He quotes Goethe (who had proposed a law of compensation or balance of growth): "in order to spend on one side, nature is forced to economise on the other side." And he then adds a few examples of his own: "It is difficult to get a cow to give much milk and to fatten readily. The same varieties of the cabbage do not yield abundant and nutritious foliage and a copious supply of oil bearing seeds." Darwin goes on to say that although this law is applicable to animals and plants under domestication, he does not believe it is applicable to species in the wild. Although he adds, "many good observers, more especially botanists, believe in its truth."

It is very likely that the botanists Darwin had in mind included his friend Joseph Hooker, director of the Royal Botanic Gardens at Kew, who would have explained to him that great commercial rewards awaited the first plant breeder to produce a black tulip or a blue rose, but despite more than two centuries of cross-breeding experiments no one had come close to producing such a variety because they had encountered the same barrier.

The unsuccessful experiments carried out by the French botanists, by the American geneticists, and by generations of hopeful Dutch tulip breeders are all concerned with exploiting the natural variability that exists in every animal or plant. All species, according to botany and zoology, exhibit subspecific variation, which is merely the scientific way of expressing the fact that all individuals are different, within certain limits.

All the different breeds of dog, for example, from the tiny Chihuahua to the Alsatian, from the Pekinese to the Great Dane, are all members of a single species—*Canis familiaris* or the common dog. In exactly the same way, all the races of humankind are members of the same species (indeed, the same subspecies, *Homo sapiens sapiens*) from the pygmies of Borneo to the Zulus of the African plains and from primitive natives to professors of biology.

The amount of natural variation available for cross-breeding is considerable. Dogs have been bred for their speed, for their ability to point at and retrieve game, for guard purposes, as lapdogs, and in dozens of directions more or less dictated by their breeders' whims. In cases of extreme selective breeding, such as the Pekinese and even the British bulldog, the variation achieved has been at the expense of other parts of the animals' anatomy. Both these breeds suffer breathing difficulties as a result of facial distortion. Racehorses are also bred selectively to produce a commercially successful animal and this process too can have an adverse effect on some offspring, often producing animals that are temperamental or too nervous to be ridden.

Stockbreeders and horticulturists who both make their living by breeding desirable strains would have been able to tell the French botanists and the American geneticists the outcome of their breeding trials in advance. Because they know from their own commercial experience that whenever a variation is artificially selected for an animal or plant, to produce some desirable characteristic, the "improvement" is gained at the expense of some other characteristic of the animal. With domestic animals, the corresponding loss is usually considered unimportant because it affects the ability of the animal to survive only if returned to the wild state. The modern dairy cow, for instance, is unable to go even a single day without being milked.

The natural limit on the amount of variation that can be induced in a species is merely the expression of the fact that nowhere in the animal or plant kingdom is there a species that is capable of the infinite biological plasticity demanded by evolution theory, capable of unlimited adaptation to different environments and different modes of life. Living organisms are systems with limited potential for change in which variation of one characteristic reacts on other characteristics, usually with unfavorable results.

This finding is of central importance because it is one that Darwinists will usually accept, having considered the evidence, but will later on simply forget all about when they are speaking of the Darwinian concept of variation and natural selection. It seems to bring out the Jekyll and Hyde in evolutionists from Darwin down to the present. Darwin withdrew his claim that bears could change into whale-like creatures, yet he continued to believe that microbes had evolved into men. Because this is a central issue, it is worth looking in a little more detail at exactly how variation came to occupy the position it does in the synthetic theory.

Darwin began to wonder about speciation among animals when he observed on the Galapagos Islands that the finches and tortoises on each island varied in detail from their counterparts on other islands in the group, while retaining a general similarity. The finches, for example, differed from island to island in terms of the size and shape of their beaks, and the type of food they lived on, but they remained finches. On one island they had strong, thick beaks for cracking nuts and seeds; on another island they had smaller beaks and fed on insects; on a third, the beak was suited to feeding on fruits and flowers. The acting governor of the Islands, an Englishman called Lawson who entertained Darwin over dinner, explained that he could identify the island from which a tortoise came by the shape of its shell: the Albemarle Island tortoise had a different shell from the Chatham tortoise, both of which varied from the tortoise on James Island.

Darwin generalized from these and similar observations that animals isolated by geography can change their characteristics over successive generations and adapt to different environments or ecological niches. The mechanism mediating this change he surmised was natural selection—the survival and breeding advantage of those individuals best adapted to their changed environment.

The idea of *natural* selection occurred to Darwin because he had long been interested in how domestic animals and plants are changed by *artificial* selection—by animal husbandry techniques practiced by stockbreeders and his fellow pigeon fanciers. The stockbreeder selects for breeding cattle which produce a high yield of milk and healthy calves, in order to gain those traits which will enable him to build up a dairy herd differing considerably from wild cattle in their milk-producing characteristics. In the same way, Darwin believed, the demands of the environment would act in the stockbreeder's role, favoring the well-adapted and weeding out the poorly adapted.

At the time, though, Darwin was unable to explain precisely what was the agent of change; for example, what it was that produced a cow yielding higher-than-average milk in the first place. And, more importantly, what specific biological agent caused adaptive changes of character to persist over successive generations. At first he thought the changes might be due to an amplification or magnification of the small natural differences that exist between all individuals of the same species. That there would naturally be a whole spectrum of beak sizes in finches, and that on an island where there were only nuts to eat the large-beaked variety would dominate. However, this explanation was scientifically unacceptable in 1859 because it was wrongly believed that such natural variations were *diluted* by breeding—not amplified.

This was one of the main reasons that Darwinism had almost been consigned to the scientific scrap heap by the beginning of the twentieth century, and it was also the reason why the rediscovery of Mendel's experiments in plant breeding rescued Darwinism so comprehensively.

Evolutionists in the early years of the twentieth century believed that, with the newly understood laws of genetics, they had a complete explanation for the mechanism of change. To use present-day terminology, the characteristics of each animal and plant are controlled by its genes, the complex chemical structures in its reproductive cells, which are passed on from generation to generation, and which carry the coded instructions that govern the development of the new embryo. Mendel had discovered that some genes dominate. For example, if you cross a short pea plant with a tall pea plant, most of the offspring will be tall. This would explain perfectly how the thick-beaked finch would come to dominate, because his thick beak (if controlled by a dominant gene) would be passed on to a majority of his offspring. Thus useful characteristics would be amplified by breeding, and the synthetic theory explains Darwin's original observation.

So far, no step in the chain of reasoning has been taken which goes beyond the data. But Darwin's successors felt that the theory as it stood at that point could be extended logically and naturally one step further—and it was a step that appeared to be not very far beyond the data. If variation and natural selection explained how a finch could change its beak shape to adapt to its island home, and how the giraffe's neck could get longer, then it also explained how one species could turn into a *completely different species*.

After this first intoxicating draft from the tankard of speculation, the newly hatched synthetic evolutionists were brought back to earth with a disillusioning jolt. As we have already seen, ordinary subspecific variation cannot be pressed into service as the mechanism of evolution for two important reasons: first, because of what Mayr calls "genetic homeostasis"—the natural barrier beyond which selective breeding will not pass; and, second, because the genetic program or recipe for whales is not contained in the *existing* genetic makeup of bears. A genetic change is needed before one can change into the other—and natural selection is not capable of initiating genetic change.

Physical characteristics are controlled by genes (or groups of genes acting in concert). Sexual reproduction ensures that each new individual receives a "new deal," since the genes of the parents are shuffled together and recombined like a pack of playing cards. Sometimes the "new hand" is very like the old one, as when a child strongly resembles one parent; sometimes it is very different. But, in every case the new deal can be drawn only from the existing pack, just as a hand at bridge must contain some hearts, or clubs, or diamonds or spades, no matter how much the pack is shuffled.

In terms of physical characteristics, what this means is that genetic recombination can give rise to variations that are within the range for each species: a finch with a beak a little bigger than before, or a cow that yields more milk than before. What it does not mean is that genetic variation of the ordinary kind is capable of explaining the appearance of entirely novel characteristics. It does not explain the appearance of a wing where before there was only an arm. For the genetic inheritance mechanism is merely one of a reshuffling and recombination of characteristics already represented in what Dobzhansky called the "gene pool" of that species.

This is the specific reason that Dutch tulip growers have never been able to achieve a black tulip or rose breeders a blue rose. There is no gene for black coloration in the gene pool of the tulip. And, sadly, there are no blue genes for the rose.

Ironically, Darwinists were rescued from this dilemma by one of the three botanists who had simultaneously rediscovered Mendel's pea experiments. Holland's Hugo de Vries found the answer in the phenomenon he believed he had discovered in the evening primrose and which he had dubbed "mutation." This was ironic because de Vries believed he was formulating a rival theory to Darwin's. Ultimately, though, de Vries' mutation was absorbed by Darwinian evolution and became the synthetic theory.

This time there seemed no reason to doubt that the theory and the data were in perfect agreement. It was spontaneous, random mutation of the genes that caused novelties to arise; Mendelian genetics that enabled these mutations to be inherited by a majority of offspring; and natural selection that ensured the dominance of the best-adapted species. The fundamental mechanism underlying evolution was therefore chance mutation, acting together with natural selection.

Mutation, as understood by synthetic evolutionists, means the spontaneous change in chemical composition of the genes which occurs quite independently of, and in addition to, the reshuffling and recombination that ordinarily occurs in sexual reproduction. Mutation is today interpreted as being a spontaneous change in the sequence of nucleotides composing the DNA molecules contained in the chromosomes or a spontaneous change in the whole chromosome—a subject looked at in detail in chapter 14. These changes can be brought about by radiation, by chemical agents, or simply by copying errors when a cell divides and the DNA double helix separates and replicates itself.

But the most important thing to be understood about mutation is that it is the *only* mechanism proposed by synthetic evolution theory that can account for the appearance of novelty in form. To repeat, ordinary genetic recombination can account for minor changes in characteristics of an individual—blue eyes rather than brown, tall rather than short—but it can never account for the appearance of any characteristic which is not already contained in the gene pool of that species.

A breeding pair of white mice might give rise to a brown mouse, providing that the genetic code for brown coloration is already present in the reproductive cells of at least one parent. But they will never, by natural means, give birth to a green mouse, since the genetic code for green coloration is not present in any strain of mouse. The only means by which any entirely novel characteristic can come into being is through the mechanism of genetic mutation—the genetic material in the reproductive cells of one parent must undergo a spontaneous change which must retain the genetic integrity of the original program and hence the viability of the offspring, but which causes some change in form.

The only way a bear can become a whale is through mutation. No amount of natural selection alone will do it, as Darwin was at first inclined to think.

But hold on a moment: What about Darwin's original observations on the voyage of the *Beagle*? What about the Galapagos finches? Are they not living proof that, when a species is isolated geographically and by habitat, its individual members will diverge imperceptibly until they are no longer one species but many? Does this not prove that speciation occurs naturally?

CHAPTER 13

The Beak of the Finch

D ARWIN'S LIVING MONUMENT on the Galapagos, the islands that were so fruitful for his thinking on evolution, is the finch that bears his name.

This little bird has come to occupy a special place in the history of Darwin's theory for it is claimed to demonstrate, in the words of writer Jonathan Weiner, "evolution in real time."[1] This dramatic claim is made because Darwinists believe that they have actually observed the process of variation and natural selection as it takes place in Darwin's finches on the islands—in a word, the process of speciation.

Speciation is a term used by Darwinists to describe a process in which one species of bird, animal, or plant evolves into two or more species. It is thus a key concept and of central importance to the neo-Darwinist theory of evolution.

Darwin's original theory offered a single mechanism for the origin of species: the natural selection of variations that exist from individual to individual. At the beginning of this century the discovery of Mendel's work on the mechanism of genetic inheritance, and the phenomenon of mutation, were married to Darwin's original conception to give the neo-Darwinist theory; that the mechanism of evolution is the natural selection not just of ordinary variations but of genetic mutations.

In both cases the theory is critically dependent on, and intimately bound up with, the idea of the "species." And before we can determine with any confidence whether Darwin's finches really do demonstrate evolution from one species to another, we must first answer the question, What is a "species"?

The word species is defined in the biological sense by the *Oxford English Dictionary* as "A group or class of animals or plants (usually constituting a subdivision of a genus) having certain common and permanent characteristics which clearly distinguish it from other groups."

This definition seems perfectly straightforward. Yet the dictionary goes on to add a very rare caveat to the definition when it says, "The exact definition of a species and the criteria by which species are to be distinguished (especially in relation to genera or varieties) have been the subject of much discussion." As we will see, this qualification is a masterpiece of understatement.

The definition used by the dictionary is the commonsense or folk definition that was adopted by biologists throughout the nineteenth century. But with the advent of a more scientific, research-based approach to biology in the present century it became clear that it was an inadequate definition. It is particularly defective when used in any context involving the discussion of evolutionary theories and mechanisms.

The most obvious immediate problem—both for biologists concerned with classifying nature, and for lexicographers concerned with classifying words—is this: if members of a species vary, how can you tell when a particular individual either is, or is not any longer, a member of the species?

For example, is a mule a horse or an ass? Or is it some kind of halfway house between the two? And are the criteria used to decide its status arbitrary criteria, or do they spring from some deep structural principle which reflects the way in which nature is truly organized? Is it a matter of scientific opinion or scientific fact? If fact, then it can be made the basis of a theory of evolution. If opinion, then it cannot.

After much debate on this subject in the first half of the twentieth century, a group of the most distinguished life scientists, including Ernst Mayr and Theodosius Dobzhansky, adopted a purely practical definition of what constitutes a species. They said, Let us define a species as a group of plants or animals that are able to interbreed and produce fertile offspring and are reproductively isolated from other such groups.

On the face of it, this definition has the advantage that it provides an empirical test of what a species is. It was not a matter of opinion among scientists; it was a matter that could be decided by experiment. As we will see later, the definition is not quite as clear-cut as it appears.

However, it was on the foundation stone of this definition that the modern neo-Darwinist theory was erected. Darwinists are able to make statements such as "the many species of finch on the Galapagos Islands are all related and have evolved from a common ancestral species" only because of the existence of a definition of "species." If there is no such definition then the statement is deprived of any scientific content. Without a definition we would be unable to say whether this statement is true or false, or, indeed, to use the words *finch* or *species* meaningfully at all. The most we could say would be that the finches, and all individual living things, are somehow related—an uninformative statement.

By the 1960s it had become clear that there were major problems with the biological definition. The very feature that seemed to be its greatest strength—its practical, empirical nature—turned out to be its greatest weakness. For in practice the definition was found not to be workable.

First, the test is not applicable to plants and animals that do not reproduce sexually such as sea squirts or self-pollinating plants. This is a substantial fraction of the biological world.

Second, the test cannot generally be applied to plants and animals that are extinct and that are known only from their fossils. Again, this is the majority of organisms. Third, the test led to some inexplicable anomalies. For example there are some breeding populations (such as of the fruit fly *Drosophila*) that are described as separate species and that do not (or cannot) interbreed, but which are genetically identical.

And, finally, there are a few known counter examples. For example, any offspring of a horse and an ass (a mule or hinny) should be infertile. Yet a few cases have been reported of hinnies bearing offspring. Similarly, the domestic bull *Bos taurus* can be crossed with a North American buffalo *Bison bison* to produce a hybrid, the cattalo, which is fertile. So the definition is a useful generalization, not a watertight criterion.

Does any of this matter? As far as biological research is concerned, it is generally of little concern to working scientists. But as far as the neo-Darwinist theory of evolution is concerned it is of crucial importance.

Suppose, for example, that a Darwinist scientist wishes to make a case for evolution from one species into another that depends on fossils. How is he or she to make that case if the definition of what constitutes the ancestor species and the descendant species cannot be biologically tested?

This is not merely an objection in principle. As we have seen earlier, mistakes of just this kind were made when fossil ammonites were described as being ancestor and descendant which today are believed to be an example of extreme sexual dimorphism—the male and female of the same species being radically different in shape. The important point to notice is that the modern interpretation is just as suspect as the original interpretation and for just the same reason. In both cases it is impossible to apply to the fossils the test of what is a species because their whole line is extinct.

Of course, in some cases, our inability to apply the definition of a species need not be a total bar to forming a reasonable judgement. For example, all living mammals reproduce sexually. It is not unreasonable to assume that fossil horses also reproduced sexually. Where the inability to apply the species test becomes a problem is when a paleontologist wishes to call one "species" of extinct quadruped the ancestor of a second "species" of extinct quadruped without being able to apply the breeding population test—as George Simpson did with his famous horse-ancestry chart.

There are thus some cases where the inability to apply the species test, or even a simple failure to apply the species test, has fatally undermined attempts to prove the existence of neo-Darwinist evolutionary processes. As we shall see, Darwin's finches are just such a case.

First let's look at what is claimed for these famous birds. In the various islands in the Galapagos group there are said to be thirteen species of finch. All these species are believed to be descendants of an ancestral finch species and to have diverged in character to inhabit the different ecological niches available in the islands, which are very remote, some 600 miles west of Ecuador, and thus provide an undisturbed natural laboratory.

Today's finches vary in their physical form (mainly the size and shape of their beaks), their habitat, and their diet, depending on which islands they inhabit. On Daphne Island, for instance, is a species called *fortis* with a strong, thick beak for cracking nuts and seeds; while on Santa Cruz Island is a cactus finch *scandens*, with a narrow fine beak, that feeds on insects.

Darwin arrived at the Galapagos in the *Beagle* in 1835. In his *Journal of Researches* (popularly known as *The Voyage of the Beagle*) Darwin famously commented that, "in the thirteen species of ground-finches, a nearly perfect gradation may be traced from a beak extraordinarily thick to one so fine that it may be compared with that of a warbler. I very much suspect that certain members of the series are confined to different islands."[2]

Darwin went on to add, "Seeing this gradation and diversity of structure in one small, intimately related group of birds one might really fancy that, from an original paucity of birds in this archipelago, one species had been taken and modified for different ends."

These tentative statements contain all the main elements of Darwinism, then and now: there are a multiplicity of "species" with a generic similarity; they live apart from each other and are "confined" to different islands; they have adapted to the differences of habitat on those islands; they represent a graded series and look as if they have all descended from a common ancestor. These facts alone invite us to draw the inevitable conclusion, without any further evidence, that the finches represent an example of evolution by natural selection. And that is precisely the conclusion Darwinists have drawn for 130 years.

Ornithologist David Lack visited the islands in 1937 and stayed through one breeding season. He built cages and tried to encourage the thirteen "species" to mate, noting that they were reluctant to mate and did so only "rarely." Lack also noted in his monograph, *Darwin's Finches*, "In no other birds are the differences between species so ill-defined."

Lack drew up maps of the islands, showing the distribution of the thirteen species. His maps showed that either one species had come to dominate each island, or two main species were in competition there.

When he returned to England with his data, apparently showing the possibility of natural selection at work, he was urged by Julian Huxley to publish as soon as possible because his work would help establish the acceptance of Darwinian processes. Lack's work was incorporated into the flourishing theory of neo-Darwinism throughout the 1950s and 1960s and Darwin's finches became as familiar to students as the melanic form of the peppered moth.

Lack was followed by the husband-and-wife team of Peter and Rosemary Grant who have lived and worked on the Galapagos Islands from 1973 to the present. The Grants and their coworkers have been the first people to study the beaks of the finches in detail. They have shown how a difference of half a millimeter in the length of a beak makes the difference between life and death for the finches during a drought.

Jonathan Weiner wrote that

> Among *fortis*, [the Grants] already knew that the biggest birds with the deepest beaks had the best equipment for big tough seeds . . . and when they totted up the statistics, they saw that during the drought, when big tough seeds were all a bird could find, these big-bodied, big-beaked birds had come through the best. The surviving *fortis* were an average of 5 to 6 per cent larger than the dead. The average *fortis* beak before the drought was 10.68 mm long and 9.42 mm deep. The average beak of the *fortis* that survived the drought was 11.07 mm long and 9.96 mm deep. Variations too small to see with the naked eye had helped make the difference between life and death. The mills of God grind exceedingly small.[3]

This is indeed convincing. Tiny, imperceptible differences in beak shape are the difference between survival and extinction. From the point of view of Darwinian evolution, however, the question is rather different: Can such differences lead to changes from one species to another?

As observed earlier, the force of these findings depends entirely upon the question of whether the thirteen species really are different species or merely variations of the same species of finch. To determine this, according to the accepted biological definition, we must find out if they mate and bear fertile young but are reproductively isolated.

David Lack tried to observe a finch of one species pairing off with another but did not find a single case. He reached the conclusion that "clearly hybridization between species is rare, if not absent." This conclusion was of crucial importance to Darwinists like Huxley because it proved that the different finches were indeed different species. And this in turn made it likely that, far from outside influences, they had diverged from a common ancestral species by natural selection in the perfect experimental setting of the Galapagos.

If Lack's observation is true, then Darwin's conjecture may also be true. If it is not true, then Darwin's idea is deprived of any content. For if all the finches on the Galapagos are merely members of the same species, then there is no meaningful sense in which they can be held up as an example of "evolution in real time."

On this key issue Jonathan Weiner seems entirely unconscious of the scientific significance of his own reporting. In his Pulitzer Prize–winning book, *The Beak of the Finch*, he wrote, "Back in 1983, for instance . . . a male cactus finch on Daphne Major, a *scandens*, courted a female *fortis*. This was a pair of truly star crossed lovers. They were not just from opposite sides of the tracks, like the Prince and the Showgirl, or from two warring families, like Romeo and Juliet: they belonged to two different species. Yet during the chaos of the great flood, they mated and produced four chicks in one brood."[4]

Not only did the finches in question mate successfully, their offspring proved to be among the *most* fertile recorded by the Grants during their twenty years on the islands. The four chicks of this mating produced no less than 46 grandchildren.

The Grants recorded many other pairings of "different species" of finch, which, like Lack before them, they dubbed "hybrids." But of course the central significance of this finding is that the identification of the thirteen varieties as different species is impossible to maintain once it is admitted that they can interbreed and produce fertile young.

The fact that different varieties prefer not to mate is very different from saying that they are unable to do so. Great Danes do not usually select toy poodles as potential mates (and vice versa) but they are capable of bearing fertile young if mated and are members of the same species, *Canis familiaris*. Arab stallions do not normally select Shetland ponies as mates, but they are members of the same species, *Equus callabus*.

Moreover, the Grants' observations undermine another myth about Darwin's finches—that individual species are "confined to certain islands." In order for different species to mate, they clearly have to occupy the same territory. Other visitors to the Galapagos have confirmed that this is this case. Television documentary filmmaker Gillian Brown spent a year working at the Darwin Research Station on the islands. It is common, says Brown, to find the different species all over the archipelago, rather than obeying the invisible boundaries of the colored territorial maps drawn up by Darwinist ornithologists.

In almost all respects, the finches of the Galapagos are so similar that it is difficult to tell them apart. Indeed, Weiner himself remarks that "some of them look so much alike that during the mating season they find it hard to tell themselves apart." This mirrors David Lack's observation that "in no other birds are the differences between species so ill-defined." The finches all have dull plumage, which varies from light brown to dark brown, all have short tails, all build nests with roofs, and all lay white eggs spotted with pink, four to a clutch.

It is very difficult for an objective observer to see how a group of finches who "find it hard to tell themselves apart," and who do in fact interbreed, can legitimately be called different species. What is the basis of this identification?

All biologists who are convinced Darwinists are perfectly well aware of the kind of problem outlined above connected with defining species. They are equally well aware of the absence of transitional species in the fossil record and of the failure of biology to find any evidence of evolutionary transitions at the species level.

The scientists who originated the biological definition of a species also appear to have been well aware of the problems they were creating for themselves. In the 1937 edition of his book *Genetics and the Origin of Species*, Dobzhansky proposed this definition of a species:

> That stage of evolutionary progress at which the once actually or potentially interbreeding array of forms becomes segregated into two or more separate arrays which are physiologically incapable of interbreeding.[5]

This is clearly a very rigorous formulation because it places the onus on the scientist to prove, by experiment, that interbreeding is actually impossible—not merely that the individuals concerned no longer prefer to mate, but that they are "physiologically incapable" of interbreeding. This would certainly be a test that was unambiguous and, in principle, infallible.

But by 1942 Dobzhansky's fellow biologist, Ernst Mayr, suggested a formulation that is considerably less rigid. Mayr defined species as:

> Groups of actually or potentially interbreeding natural populations which are reproductively isolated from other such groups.[6]

By the third edition of his book, in 1951, Dobzhansky had evidently been prevailed upon by his colleagues because he had relaxed his definition to the point that it agreed with Mayr's.[7] And this is essentially the definition that was adopted and is still used today by many members of the biological community.

The important point to notice about the new biological definition of a species is that it no longer insists on determining experimentally whether the creatures in question *can* interbreed. It is enough that, for whatever reason, they do not do so. It is hardly surprising to find, then, that Darwinist biologists feel free to describe the Galapagos finches as individual species on the basis that they usually choose not to mate and to describe the occasions when they do interbreed as "hybridization." But it is from this kind of wordplay that all their subsequent claims of speciation and "evolution" flow.

The claim that speciation is an observed fact and can be evidenced by numerous examples continues to be asserted by Darwinists. In reality every one of these examples falls into one of two categories of pseudo-speciation. The first is where speciation is claimed by the same kind of semantic ruse employed in the case of the Galapagos finches— quietly dropping the rigorous biological definition of what constitutes a species and substituting a definition so ill defined that any subspecific variation can be claimed as "speciation." The second is the case where freak degenerative mutations make a new offspring genetically incompatible with its parents. This sometimes happens with plants where the number of chromosomes doubles, creating a "sport" (it was such an anomaly that de Vries observed in the evening primrose). Whether this is described as speciation or not, neither case can ever give rise to evolution in the true Darwinian sense: a mechanism that could explain the transformation of a bear into a whale.

The solution modern Darwinists have adopted to these problems is breathtakingly simple. First they have drawn a distinction between macroevolution and microevolution.

Macroevolution, they say, is the new name for Darwinian speciation, the process by which species (organisms so genetically different they can no longer interbreed) come into being. This process occurs over millions of years so it cannot be observed or made the subject of experiment.

Microevolution, on the other hand, is very much simpler. It is the change in frequency of variant genes (called *alleles*) from generation to generation, and something that can be observed. Darwin's finches are an example of microevolution. By defining microevolution in such simple terms, Darwinists are sure of silencing any critics, for no one can disagree that variant genes do change in frequency from generation to generation, just as no one can disagree that a bird with a thick beak is genetically different from a bird with a thin beak.

It is the next part of the argument (where the goalposts are moved) that is the really clever part.

When you get enough microevolution, say Darwinists, you eventually get macroevolution. This proposition cannot be tested empirically for exactly the same reasons that the concept of macroevolution itself cannot be tested experimentally. Once you have agreed with the first part of this proposition, however, it appears difficult not to agree with this final part.

In fact, this final proposition—that lots of microevolution adds up to one big macroevolution—is contradicted by every objection raised against neo-Darwinism in the past fifty years: that what Mayr called genetic homeostasis will prevent morphological change beyond a certain point; that there is no evidence for gradual change leading to macroevolution in the fossil record; that billions of years are required to accumulate such microevolution; and so on and so on.

Darwin saw natural selection by itself not as the mechanism of evolution in the sense of mediating the change of one species into another species, but as a mechanism that can lead to such macroscopic evolution. That is not how modern Darwinists see things, though. The modern view is that natural selection is responsible for selecting which variant genes are passed on from generation to generation ("microevolution"). Because in this modern view macroevolution is simply accumulated microevolution, then this process is evolution itself. Thus modern biologists have taken a step which Darwin never would take: they have directly equated natural selection with evolution itself.

Above all, the objection to Darwin's finches as evidence for evolution is that—as we saw in the previous chapter—just as Darwin's bear can become a whale only by mutation, not by natural selection, so Darwin's finches, and every other living thing, can become a different species only by undergoing a spontaneous genetic mutation.

"It must not be forgotten," says Ernst Mayr, "that mutation is the ultimate source of all genetic variation found in natural populations and the only new material available for natural selection to work on."[8]

We must also not forget that the words "natural selection," if they mean anything, must mean "choosing the one or the few from the many." To select is to pick from a larger number. Thus whatever else "natural selection" may be it is inescapably a mechanism that *reduces* biological diversity. At the same time, it is clear that Darwinian evolution is a process that essentially involves the *increase* of biological diversity—the origin of species, in fact, not their reduction.

In seeking the creative engine of evolution in Darwinian theory, we must abandon "natural selection" and turn instead to the other part of its twin mechanism, the phenomenon of "spontaneous genetic mutation." It is to this mysterious matter that we now turn.

CHAPTER 14

Of Cabbages and Kings

T HE INSTRUCTIONS FOR REPRODUCTION of a cabbage or a king are contained in dark threadlike strands in the nucleus of its cells, called chromosomes, which are long molecules of DNA sheathed in protein. The instructions themselves consist of sequences of four chemical groups, called nucleotides, strung like clothes on a washing line and identified by their initial letters—C, T, A, and G (cytosine, thymine, adenine, and guanine).

The genetic meaning of each sequence—the kind of physical characteristic it gives its owner—depends on the sequence of nucleotides (C, T, A, and G) and the position of that sequence on the "washing line." The sequence TCA for example is the genetic code for an amino acid called serine which is important in building membranes. The sequence CCA is the code which causes the synthesis of an amino acid called proline which is widely used in building connective tissues.

Each chromosome is like a magnetic tape. Not surprisingly, the total number of instructions, or sequences, in an organism as complex as a human runs into millions—too many in fact for one tape. They run over onto another tape, and another. In humans there are 23 chromosomes in the "tape library." In the fruit fly there are 4, while the humble land snail has 27 and, curiously, the simple goldfish has 47. When she wrote that a rose is a rose is a rose, Gertrude Stein must have been unaware that roses exist with 14, 21, 28, 35, and 56 chromosomes.

Despite the genetic code taking up so much room (actually a chromosome is only one-hundredth of a millimeter long) some 90 percent of the recording space in the tape library is empty. The genetic instructions which actually cause the manufacture of proteins occupy only 10 percent of the available coding space. Or to be more exact, only 10 percent of the sequences in the chromosomes cause anything to be replicated. The function of the other 90 percent of sequences is unknown at present. Although it does not replicate, it may affect the positioning (and hence the genetic meaning) of the sections that do replicate.

While enough is known of the genetic code to justify saying that geneticists understand it in principle, much of it remains mysterious. Unfortunately, it is not a simple case of a single gene, at a single location, controlling a single characteristic. Various locations are linked together to control groups of characteristics in a non-obvious way.

In sexual reproduction, when a male sperm unites with a female egg, the DNA molecules of each split apart, like a zipper unzipping, and cross-join to form the set of chromosomes of the first cell of the new individual. Also, when the first cell divides and redivides to form the new embryo, the DNA molecules unzip and replicate themselves exactly, thus perpetuating the program or building instructions in every cell of the new individual both for its embryonic form and for life.

A number of things can go wrong with genetic reproduction. Individual genes (nucleotide sequences) can sometimes be replicated incorrectly, either because of faulty selection of nucleotides or because the right sequence is put in the wrong location, perhaps shifted one position to the left or right. Faulty replication of a sequence will result in the issuing of instructions to the factory to make the wrong product. If the sequence TCA is wrongly replicated as CCA (just like a typing error in a document) then the cell will manufacture proline instead of serine, with unpredictable consequences.

The Darwinist interpretation of these discoveries is that the genetic mutation which causes novelties in form is caused by spontaneous alterations in the DNA molecule and hence of the genetic code. A good many of these genetic changes happen without actually altering the physical characteristics of the individual who carries them. These genes are called inert, latent, or inhibited. It is believed by Darwinists that they are "stored" in the 90 percent of unused genetic material. Under certain circumstances these inert genes can replace the normal genes and become expressed in the physical characteristics of the offspring.

Evolution, according to Darwinists, is due basically to copying errors. Although the DNA is astonishingly stable from generation to generation, and although reproduction is error free to a far higher degree than the most efficient man-made copying systems, there are occasional mistakes: an A appears where a T should be, or a G appears instead of a C, or a nucleotide may be strung in the wrong position on the molecule.

These copying errors can happen spontaneously, or they can be caused by some outside mutagenic agency such as radiation or highly toxic chemicals, like mustard gas. Ultraviolet light from the Sun is mutagenic, but has very little penetrating power and hardly gets beyond the skin. On the other hand, X rays penetrate deep into the human body causing considerable direct cell damage and damaging DNA which will begin to replicate in a faulty way.

The results of such copying errors are tragically familiar. In body cells, faulty replication shows itself as cancer. Sunlight's mutagenic power causes skin cancer; the cigarette's mutagenic power causes lung cancer. In sexual cells, faulty reproduction of whole chromosome number 21 results in a child with Down's syndrome. Only "germinal" mutation—that is, the mutation of sexual cells in the male sperm or female egg—can result in an inheritable variation, believe Darwinists.

According to the same theory, "somatic" or body-cell mutation cannot be inherited, and this is the specific reason that Darwinists are also anti-Lamarckian. They believe that even if an animal's mode of life should result in somehow bringing about mutation in the creature's body cells, there is no mechanism for these changes to be passed on to the next generation; only reproductive cells do that job, not body cells.

In trying to assess whether Darwinists have made their case, the key issue in molecular biology is the rate of mutation. This has to be frequent enough to provide a realistically probable occurrence of novelties, but not so frequent that no two generations are ever the same and evolution runs haywire. Too little and we are stuck in the primeval ocean unable to set the evolutionary ball rolling; too much and we are living in an unstable nightmare world of monsters.

The question of just how much mutation takes place is of considerable importance, and is a much studied subject. What conclusions have Darwinists come to?

Julian Huxley estimated that the rate of inheritable mutation was around one in every million births.[1] French biologist Jacques Monod has estimated the rate at one in ten thousand births.[2] The reason for this diversity of opinion between the professor of zoology at King's College, London, and the director of Paris's Pasteur Institute is simple: it is because the beneficial spontaneous genetic mutation remains no more than a hypothetical necessity to the neo-Darwinist theory.

No one has ever observed a spontaneous inheritable genetic mutation that resulted in a changed physical characteristic, aside, that is, from a small group of well-known and usually fatal genetic defects. Because no one has ever observed such an event, no one really knows whether they occur at all and, if so, how often. Because deleterious mutations are known to occur, Darwinists appeal to the statistics of large numbers. If deleterious mutations can occur, then given enough time beneficial mutations can occur. There is no evidence for this claim. But it is irrefutable.

"Detectable results of germinal mutation among people are only very rarely encountered," says the author of the 1984 *Encyclopaedia Britannica* entry on Human Genetics; "thus the actual rate of mutation in human chromosomes defies full measurement. Efforts to measure mutation rate therefore are most conveniently directed towards selected dominant . . . mutations for which [physical] recognition is easier; indirect (inferential) methods of measurement are still required."[3]

The dominant mutations where physical recognition is easier include achondroplasia (dwarfism), Huntington's chorea, and Down syndrome. And it is from deleterious mutations of this kind that the human mutation rate is estimated. *Encyclopaedia Britannica* cites a general mutation rate for human genes of 4 mutations per 100,000 gametes (that is, male sperm or female egg).

This rate of mutation sounds impressively high. But the reality is very different. What has happened is that the rate of mutation has been inflated by the simple device of making the definition of the term "mutation" so elastic that it can include any and every inheritable change—including those that invariably lead to fatal diseases. Even though Darwinists are well aware that no individual can ever benefit from dwarfism, they include achondroplasia as a genetic mutation; they also include Huntington's chorea, and neurofibromatosis, and use them all to measure the rate of mutation.

In addition, Darwinists lump in other types of mutation, many of which are caused by natural radiation and other mutagens, and which again can only result in exclusively harmful results for the offspring. It is obvious that, in reality, a major proportion (perhaps all) of these so-called mutations cannot possibly lead to any benefit and are bound to lead only to degeneration. Only a tiny fraction have any prospect of turning into a novelty of form that might conceivably be helpful, and even the existence of this tiny minority is granted only because the claim that they might exist cannot be refuted.

The fact is that more than 99 percent of so-called mutations should not be included in the measured rate. But these mutations cannot be positively excluded because no one can predict which mutations will be useful and which will not. It is conceivable, after all, that radiation damage might produce a useful variation, or that dwarfism might be adaptive if the circumstances of the environment changed.

I believe that Darwinists should muster the courage to come clean by separating the two cases, reserving the term mutation for any change in genetic coding, whatever the cause and whatever the effect, and use some other term—perhaps "novation" (novelty-producing mutation)—to describe the kind of mutation they say is potentially useful. Note that "novations" do not have to exclude all copying errors or blunders—only those whose genetic consequences are already known not to lead to evolutionary novelty, such as Huntington's chorea.

Can we estimate the rate of "novation" as opposed to the rate of mutation? Yes, we can. The rate of novation is a number that is vanishingly small (if not actually zero). It is a number so small that in order to account for synthetic evolution by random mutation, one has to have an almost religious faith in the power of extremely unlikely events and very long time scales.

Thus at the very heart of the synthetic theory of evolution is a single, central matter: improbability. How we deal with this issue alone either convinces us of the validity of neo-Darwinism or convinces us of its impossibility. Regardless of the evidence from all other sources—geology, stratigraphy, paleontology, comparative anatomy, zoology, botany, and genetic studies—it is the question of the probability of life spontaneously coming into being and spontaneously evolving—without outside assistance—that separates the sheep from the goats.

The two camps might justly be represented by their respective champions: William Paley, the eighteenth-century Archdeacon of Carlisle, and Ronald Fisher, founder of the modern mathematical school of genetic studies, both of whom are almost always quoted when the issue of spontaneous random mutation is debated.

Paley, in his influential book *Natural Theology* published in 1828, observes that if, while out walking, you were to find lying on the ground a watch full of intricate mechanisms you would have to conclude that it had been wrought by a creator; it would be impossible to believe that such a machine had come into being accidentally. The human body is infinitely more complex and intricate than any watch mechanism, so we must conclude that it too has a purposeful creator.

Fisher, whose equally influential book *The Genetical Theory of Natural Selection* appeared in 1930, observed quite simply that natural selection is a mechanism for generating improbability.

Paley's watch—the argument from design—is not really a serious scientific argument and can be easily refuted. Most recently it has been very effectively dismissed by zoologist Richard Dawkins. In his book *The Blind Watchmaker* Dawkins points out simply that we are not obliged to see the hand of God in such seeming miracles and that one individual's inability to conceive of highly improbable events does not make those events impossible. Those who employ Paley's argument, says Dawkins, should speak only for themselves.[4]

Dawkins's own way of dealing with the improbability of evolution by mutation (a way that is representative of the modern neo-Darwinist view), however, makes use of a fallacy many times more subtle than anything Paley dreamed of on his horological excursions. One has to observe each step of Dawkins's argument very carefully to spot exactly where the fallacy comes in.

To get from a barren primeval Earth to a complex organ like the human eye in a single step, says Dawkins, would require random spontaneous events that are so improbable as to be practically impossible. However, he says, it is not so wildly improbable to get there in a series of small steps, each step requiring admittedly improbable events, but events not so improbable as to be practically impossible. And, of course, vast ages are available in the geological past for these smaller steps to be accumulated. Adding up a long series of small, improbable—but not impossible—steps can cumulatively give rise to a complex mechanism such as the eye which *overall* is of incredible improbability.

More simply, you can get a result whose improbability is so great as to be practically impossible, by adding together a lot of little steps whose improbability is high, but nevertheless practically possible. Moreover, says Dawkins, if you break up the process into steps that are cumulative, it is quite likely that you only have to contend with one or a few steps of extremely low probability—those at the beginning—and once the evolutionary ball is rolling, the events required become less and less improbable.

"My personal feeling," he says,

> is that, once cumulative selection has got itself properly started, we need to postulate only a relatively small amount of luck in the subsequent evolution of life and intelligence. Cumulative selection, once it has begun, seems to me powerful enough to make the evolution of intelligence probable if not inevitable. This means that we can, if we want to, spend virtually our entire ration of postulatable luck in one big throw, in our theory of the origin of life on a planet.[5]

Dawkins's argument is a modern rendition of the traditional Darwinist approach and the error it falls into is that dubbed the "statistical fallacy" by Francis Crick.[6] Although employing modern concepts, it is really in principle the same as Darwin's own claim that,

> Slow though the process of selection may be, if feeble man
> can do so much by his powers of artificial selection, I can
> see no limit to the amount of change, to the beauty and
> infinite complexity of the co-adaptations between all or-
> ganic beings, one with another and with their physical con-
> ditions of life, which may be effected in the long course of
> time by nature's power of selection.

If Paley's watch is the argument from design, then the Darwinian case might be called the argument from probability. What does it really amount to?

Suppose we have a highly improbable event such as a perfect deal in bridge, where each of the four players receives a complete suit of cards. The odds against this happening are billions of billions of billions to one. Let us assume that, since being manufactured, the cards have been used for 99 deals and on the 100th time the pack was shuffled the perfect deal arose. Can we say that each of these previous shuffles, deals, and plays of hands (number 1 for instance) was a cumulative event that ultimately contributed to the perfect deal? Can we reduce the ultimate odds against the perfect deal by attempting to spread them around more thinly between the intermediate steps? Not *afterward*, when we know the result, but at the time each step is occurring?

The answer is no, we cannot. Like the supposedly evolving DNA, the cards have a memory in that the previous deals have contributed to their current order and the ultimate perfect deal. But being partway toward a perfect deal does not alter the odds on the ultimate deal, because some of the key random events determining the ultimate outcome *have not yet taken place*.

The same is true of Dawkins's hypothetical evolutionary model. Although the earlier steps in his evolution process are seen retrospectively to contribute to the end result, that does not affect the probability of each intermediate step coming about *at the time*. It is certainly true that the minimum overall probability we have to deal with, when considering the evolution of a human eye, is a product of all the probabilities of the individual steps necessary to attaining that end. But paradoxically, this does not diminish the probability of each individual step when the need for the correct sequence is also taken into account.

What Dawkins is saying with his cumulative evolution argument is that the probability of each single step in a cumulative process must be *less* than the whole probability of leaping straight to the end result, simply because each step itself is less than the whole. But this is simply wrong. The improbability of step number 2 correctly following step number 1, correctly followed by step number 3 and so on for 100 mutations, is as great as leaping to the 100th step in one go.

What is more, the greater the number of steps into which we break up the overall leap, the more improbable it becomes that they will all take place in the right order. Mutation number 1 might be the first step in evolving an eye (or magnetic or infrared or X-ray detector). But the probability of the next mutation step affecting that organ being the second step needed for an eye is not increased thereby. It does not become any easier for an eye to come into being just because the first of the 100 or 1,000 accidents needed has taken place, *even if that first step is a very important general innovation such as light-sensitive tissue.*

Modern Darwinists seem to have a profoundly optimistic belief that the occurrence at an early stage in evolution of such a fundamental innovation—cells which are sensitive to light—makes cumulative selection of vision somehow less improbable. But the existence of light sensitive tissue has no effect whatever on the probability of the mutation of a lens, or an iris mechanism or an eyelid or anything else.

Of the vast range of characteristics spelled out by DNA, the next copying error is more likely to be about something else entirely—the beginnings of a wing or lung perhaps—or it may be the wrong step, such as providing eyelids before providing the muscles to move them, thus blinding their possessor.

Darwinists say that this case cannot be made against them because purpose has no place in their argument. The Darwinist mechanism of evolution is blind; and its outcome is arbitrary. Neither nature nor Darwinists care what the end result of the selection process is: the species that inherit the earth will simply be those that nature has blindly selected and that are best adapted to their habitats and way of life. There is no "perfect deal" in evolution, say

Darwinists; no final result to be anticipated. There is only an infinity of uncertainty leading on always to novelty dictated only by changed environmental circumstances.

This apparent rejection of purpose is deceptive. Darwinists are very firmly convinced that there is a predictable, and desirable, end result for any given habitat or way of life. Indeed, that is the very origin of the Darwinian concept. Any species less than perfectly adapted will ultimately be replaced by another species that is better adapted, unless it has no competitors, in which case it will ultimately acquire such competitors and be displaced. This process will inevitably continue until eventually it will reach a conclusion where the process cannot take place any further.

Although the mutation part of this process is random it is clear that the selection part most certainly is not random: in fact it is keyed perfectly to the template of ecology and habitat. Otherwise why would camels be able to survive for days without water? Or sea otters hold their breath for long periods?

Darwinists use this very argument to account for the parallel evolution, in isolated environments, of identical animals, such as Tasmanian marsupial wolves and European placental wolves.

Given any specific habitat, and any existing set of animal characteristics, it is possible in principle to set down precisely what characteristics that animal will have to acquire to become perfectly adapted—and thus irreplaceable. In theory it ought to be possible to use one of the Darwinists' favorite computer-based genetic software systems to show in live animation exactly how that creature should mutate to get most efficiently from its current position to the theoretically perfect position.

This idea is not merely conjecture. Darwinists have already done this very trick in the case of the extinct dinosaur. In 1982 Dale Russell and R. Seguin of Ottawa University published a paper describing the partial skeleton of the dinosaur *Stenonychosaurus* which had been found in Alberta in 1967. Their paper covered the work involved in building a flesh-and-bone reconstruction of this species. However, Russell and Seguin decided to take their reconstruction one step further. Because *Stenonychosaurus* was small (about 6.5 feet tall) and a biped with flexible fingers and a relatively large brain, the authors asked what would have happened if, instead of becoming extinct, the creature had continued evolving in a Darwinian fashion to the present day. Their resulting reconstruction looks astoundingly humanlike apart from a certain unfortunate reptilian glare.[7]

Russell and Seguin's reconstruction is, of course, merely an imaginative thought experiment, carried through to be entertaining and thought-provoking. But I believe it represents quite fairly the belief shared by most people—Darwinists and non-Darwinists alike—that there is an inevitability about the design of man and of all other species. There is a beauty and grace in the flight of the bird which less efficient flying designs do not possess and that enables birds not only to conquer the air but to dominate it.

This perfect fitness of form is also evident in the improving design of human artifacts such as the car and the jet airliner: decades of experience with a plurality of trial designs passing through the filter of experience into a single optimum design—a process frequently referred to as design evolution (though not, of course, happening by chance).

Darwinists should steel themselves to recognize that the flight of the eagle and the sprint of the cheetah represent a "perfect deal" for evolution, an end result that is the best of breed. These animals have not arrived at an arbitrary point in genetic space; they have arrived at the point that uniquely positions them to best exploit their habitats.

Throughout the lifetime of the Earth, there always was a definable probability of these animals getting to that end point by random mutation—and that probability always was vanishingly small.

Simpson's claim, quoted in chapter 11—"The characteristics themselves do not directly matter at all. All that matters is who leaves more descendants over the generations. Natural selection favors fitness only if you define fitness as leaving more descendants"—simply will not do. It is an excuse Darwinists hold in reserve in case they are asked to comment analytically about the inherited characteristics of any given animal or plant.

Of all the difficulties facing neo-Darwinism, the improbability of spontaneous genetic mutation leading to beneficial novelties in form ought to be the major source of concern. This is so because, as explained in the previous chapters, it is the one and only source of inheritable variation available above the species level—the ordinary variation caused by genetic recombination not being capable of producing novelties above the species level.

Ronald Fisher's often quoted observation that "natural selection is a mechanism for generating improbability" can now be seen to be both illogical and irrelevant to the debate on evolutionary processes. Improbability has nothing to do with natural selection: it is connected entirely with the genetic mutation part of the Darwinian mechanism. There is nothing improbable about a dark moth surviving on a dark tree while a light moth is eaten; there is only something improbable about a melanic mutation occurring, purely at random, in a species for whom it is indispensably necessary for survival.

In practice Darwinists get around this problem in a number of ways, two of which we have already seen: they pretend many more mutations occur than actually take place, by including fatal genetic defects; and they pretend that splitting up the overall evolution process of a complex organ, like the eye, somehow reduces the improbability of those separate steps coming about by accident in the correct sequence.

In addition there are two further important devices sometimes used by Darwinists that can be found in a variety of guises. The first is to ignore the difficulties inherent in genetic mutation and "fudge" it together with ordinary genetic variation—ignoring also the fact that genetic recombination alone cannot give rise to novelties in form above the species level. This is the "industrial melanism in moths" fudge, for instance. A second device is to reintroduce purpose or direction into nature by the backdoor. Darwinists often conveniently forget that chance is blind and lapse into using phrases like "selection-pressure" (a favorite phrase of both Mayr and Dobzhansky) imagining that natural selection can place an order with random mutation like diners choosing from a restaurant menu.

An unusually clear example of the corrupted vulgate version of neo-Darwinism in practice occurred with the recent broadcast of an Open University educational program on British television. The broadcast concerned certain species of wildflower that had adapted to life on railway cuttings dug 100 years ago, in rocks containing highly toxic minerals, such as arsenic, antimony, and lead. The program's presenter explained to Open University biology students that here was an example of natural selection and evolution in action. The cuttings had been dug in rocks where no flower could survive, he said, placing an extraordinary environmental demand on nature. Yet, within 100 years, species of wildflower had evolved which were able to tolerate and to thrive in the highly toxic conditions where all normal varieties had withered away.

The magnitude of this claim and the magnitude of its falseness is simply breathtaking. First, no specific change has occurred: no new species have come into existence, so the claim that evolution has occurred is simply untrue. Second, the claim that the appearance of plants which are toxin-resistant is an example of natural selection is equally false. What has happened is precisely the same as in the case of so-called "industrial melanism" in moths: the plants unable to tolerate toxic soils all died, leaving the ground clear for plants which are not poisoned by the metals in question. To imagine that a new species of plant came into existence because workmen dug over the ground is reminiscent of the eighteenth-century belief that maggots in cheese represented the spontaneous generation of life. The presenter also said that the appearance of these "novel" types was *in response* to the evolutionary demand of natural selection for just such a plant.

This kind of thinking is symptomatic of the confusion that the teaching of neo-Darwinism leads to. Though many lecturers and teachers are sufficiently well informed to know that something is amiss, they quiet their consciences with the reflection that what they are passing on to their students is merely the "popular" form of the theory, which in its true form remains inviolate and inviolable.

CHAPTER 15

The Ghost in the Machine

RUSSELL AND SEGUIN'S 1982 PICTURE of a human-looking, "evolved" version of a dinosaur was an impressive feat combining science and imagination in a constructive and entertaining way. Yet few in 1982 foresaw that, in little more than a decade, over 100 million people around the world would pay to be scared by the even more impressive feat of the computer-generated dinosaurs of *Jurassic Park*.

Nothing that has entered the evolution debate since Darwin's time has promised to illuminate the subject so much as the modern computer and its apparently limitless ability to represent, on the screen, compelling visual solutions to the most abstruse mathematical questions.

The information-handling capacity of electronic data processing, with its obvious analogy to DNA, has been enthusiastically enlisted by computer-literate Darwinists as offering powerful evidence for their theory; while genetic software systems, said to emulate the processes of genetic mutation and natural selection at speeds high enough to make the process visible, have become a feature of most up-to-date biology laboratories.

The computer has been put to many ingenious uses in the service of Darwinist theory. And it has changed the minds of not a few skeptics by its powerful visual imagery and uncanny ability to bring extinct creatures—or even creatures that never lived—to life in front of us. But, compelling though the visual images are, how much confidence should we put in the computer as a guide to the evolution of life?

In his book *The Blind Watchmaker* Richard Dawkins describes a computer program he wrote which randomly generates symmetrical figures from dots and lines. These figures, to a human eye, have a resemblance to a variety of objects. Dawkins gives some of them insect and animal names, such as bat, spider, fox, or caddis fly. Others he gives names like lunar lander, precision balance, spitfire, lamp, and crossed sabers.

Dawkins calls these creations "biomorphs," meaning life shapes or living shapes, a term he borrows from fellow zoologist Desmond Morris. He also feels very strongly that in using a computer program to create them he is in some way simulating evolution itself. His approach can be understood from this extract:

> Nothing in my biologist's intuition, nothing in my 20 years experience of programming computers, and nothing in my wildest dreams, prepared me for what actually emerged on the screen. I can't remember exactly when in the sequence it first began to dawn on me that an evolved resemblance to something like an insect was possible. With a wild surmise, I began to breed generation after generation, from whichever child looked most like an insect. My incredulity grew in parallel with the evolving resemblance. . . . Admittedly they have eight legs like a spider, instead of six like an insect, but even so! I still cannot conceal from you my feeling of exultation as I first watched these exquisite creatures emerging before my eyes.[1]

Dawkins not only calls his computer drawings "biomorphs," he gives some of them the names of living creatures. He also refers to them as "quasi-biological" forms and in a moment of excitement calls them "exquisite creatures." He plainly believes that in some way they correspond to the real world of living animals and insects. But they do not correspond *in any way at all* with living things, except in the purely trivial way that he sees some resemblance in their shapes. The only thing about the biomorphs that is biological is Richard Dawkins, their creator. As far as the "spitfire" and the "lunar lander" are concerned there is not even a fancied biological resemblance.

The program Dawkins wrote and the computer he used have no analog at all in the real biological world. Indeed, if he set out to create an experiment that simulates evolution, he has only succeeded in making one that simulates special creation, with himself in the omnipotent role.

His program is not a true representation of random mutation coupled with natural selection. On the contrary it is dependent on artificial selection in which he controls the rate of occurrence of mutations. Despite Dawkins's own imaginative interpretations, and even with the deck stacked in his favor, his biomorphs show no real novelty arising. There are no cases of bears turning into whales.

There is also no failure in his program: his biomorphs are not subject to fatal consequences of degenerate mutations like real living things. And, most important of all, he chooses which are the lucky individuals to receive the next mutation—it is not decided by fate—and of course he chooses the most promising ones ("I began to breed . . . from whichever child looked most like an insect"). That is why they have ended up looking like recognizable images from his memory. If his mutations really occurred randomly, as in the real world, Dawkins would still be sitting in front of his screen watching a small dot and waiting for it to do something.

Above all, his computer experiment falsifies the most important central claim of mechanistic Darwinian thinking, that, through natural processes, living things could come into being without any precursor. What Dawkins has shown is that, if you want to start the evolutionary ball rolling, you need some form of design to take a hand in the proceedings, just as he himself had to sit down and program his computer.

In fact, his experiment shows very much the same sort of results that fieldwork in biology and zoology has shown for the past hundred years: there is no evidence for beneficial spontaneous genetic mutation; there is no evidence for natural selection (except as an empty tautology); there is no evidence for either as significant evolutionary mechanisms. There is only evidence of an unquenchable optimism among Darwinists that, given enough time, anything can happen—the argument from probability.

But although Dawkins's program does not qualify as a simulation of random genetic mutation coupled with natural selection, it does highlight at least one very important way in which computer programs resemble genetic processes. Each instruction in a program must be carefully considered by the programmer as to both its immediate effect on the computer hardware and its effects on other parts of the program. The letters and numbers which the programmer uses to write the instructions have to be written down with absolute precision with regard to the vocabulary and syntax of the programming language he uses in order for the computer system to function at all. Even the most trivial error can lead to a complete malfunction. In 1977, for example, an attempt by NASA to launch a weather satellite from Cape Canaveral ended in disaster when the launch vehicle went off course shortly after takeoff and had to be destroyed. Subsequent investigation by NASA engineers found that the accident was caused by failure of the onboard computer guidance system—because a single comma had been misplaced in the guidance program.

Anyone who has programmed a computer to perform the simplest task in the simplest language—Basic for instance—will understand the problem. If you make the simplest error in syntax, misplacing a letter, a punctuation mark or even a space, the program will not run at all.

In just the same way, each nucleotide has to be "written" in precisely the correct order and in precisely the correct location in the DNA molecule for the offspring to remain viable, and, as described earlier, major functional disorders in humans, animals, and plants are caused by the loss or displacement of a single DNA molecule, or even a single nucleotide within that molecule.

In order to simulate neo-Darwinist evolution on his computer, it is not necessary for Dawkins to devise complex programs that seek to simulate insect life. All he has to do is to write a program containing a large number of instructions (3000 million instructions if he wishes to simulate human DNA) that continually regenerates its own program code, but randomly interferes with the code in trivial ways, such as transposing, shifting, or missing characters. (The system must be set to restart itself after each fatal "birth.")

The result of this experiment would be positive if the system ever develops a novel function that was not present in the original programming. One way of defining "novelty" would be to design the program so that, initially, its sole function was to replicate itself (a computer virus). A novel function would then be anything other than mere reproduction. In practice, however, I do not expect the difficulty of defining what constitutes a novelty to pose any problem. It is extremely improbable that Dawkins's program will ever work again after the first generation, just as in real life, mutations cause genetic defects, not improvements.

Outside of the academic world there are a number of important commercial applications based on computer simulations that deserve to be seriously examined. A good example of this is in the field of aircraft wing design where computers have been used by aircraft engineers to develop the optimum airfoil profile. In the past wing design has been based largely on repetitive trial and error methods. A hypothetical wing shape is drawn up; a physical model is made and is aerodynamically tested in the wind tunnel. Often the results of such an empirical design approach are predictable: lengthening the upper wing curve, in relation to the lower, generally increases the upward thrust obtained. But sometimes results are very unpredictable, as when complex patterns of turbulence combine at the trailing edge to produce drag, which lowers wing efficiency, and causes destructive vibration.

Engineers at Boeing Aircraft tried a new approach. They created a computer model which was able to "mutate" a primitive wing shape at random—to stretch it here or shrink it there. They also fed into the model rules that would enable the computer to simulate testing the resulting design in a computerized version of the "wind tunnel"—the rules of aerodynamics.

The engineers say this process has resulted in obtaining wing designs offering maximum thrust and minimum drag and turbulence more quickly than before and without any human intervention once the process has been set in motion.

Designers have made great savings in time compared with previous methods and the success of the computer in this field has given rise to a new breed of application dubbed "genetic software." Indeed, on the face of it, the system is acting in a Darwinian manner. The computer (an inanimate object) has produced an original and intelligent design (comparable, say, with a natural structure such as a bird's wing) by random mutation of shape combined with selection according to rules that come from the natural world—the laws of aerodynamics. If the computer can do this in the laboratory in a few hours or days, what could nature not achieve in millions of years?

The fallacies on which this case is constructed are not very profound but they do need to be nailed down. In a recently published popular primer on molecular biology, Andrew Scott's *Vital Principles*, this very example is given under the heading "the creativity of evolution." The process itself is called "computer generated evolution" as though it were analogous to an established natural process of mutation and selection.[2]

The most important fallacy in this argument is the idea that somehow a result has occurred which is independent of, or in some way beyond, the engineers, who merely started the machine by pressing a button. Of course, the fact is that a human agency has designed and built the computer and programmed it to perform the task at hand. As with the previous experiment, this begs the only important question in evolution theory: Could complex structures have arisen spontaneously by random natural processes *without any precursor*? Like all other computer simulation experiments, this one actually makes a reasonable case for special creation—or some form of vitalist-directed design—because it specifically requires a creator to build the computer and devise and implement the program in the first place.

However, there are other important fallacies too. The only reason that the Boeing engineers are able to take the design produced on paper by their computer and translate that design into an aircraft that flies, is because they are employing an immense body of knowledge—not possessed by the computer—regarding the properties of materials from which the aircraft will be made and the manufacturing processes that will be used to make it. The computer's wing is merely an outline on paper, an idea; it is of no more significance to aviation than a wave outline on the beach or a wind outline in the desert. The real wing has to actually fly in the air with real passengers. The decisive events that make that idea into a reality are a long, complex sequence of human operations and judgments that involve not only the shaping and fastening of metal for wings but also the design and manufacture of airframes and jet engines. These additional complexities are beyond the capacity of the computer, not merely in practice but in principle, because computers cannot even make a cup of coffee, let alone an airliner, without being instructed every step of the way.

In order for a physical structure like an aircraft wing to evolve by spontaneous random means, it is necessary for natural selection to do far more than select an optimum shape. It must also select the correct materials, the correct manufacturing methods (to avoid failure in service) and the correct method of integrating the new structure into its host creature. These operations involve genetic engineering principles which are presently unknown. And because they are unknown by us, they cannot be programmed into a computer.

There is also an important practical reason why the computer simulation is not relevant to synthetic evolution: because an aircraft wing differs from a natural wing in a fundamental way. The aircraft wing is passive, since the forward movement of the aircraft is derived from an engine. A natural wing like a bird's, however, has to provide upthrust and the forward motion necessary to generate that lift making it a complex, articulated, active mechanism. The engineering design problem of evolving a passive wing is merely a repetitive mechanical task—that is why it is suitable for computerization. So far, no one has suggested programming a computer to design a bird's wing by random mutation because the suggestion would be seen as ludicrous. Even if all of the world's computers were harnessed together, they would be unable to take even the most elementary steps needed to design a bird's wing unless they were told in advance what they were aiming at and how to get there.

If computers are no use to evolutionists as models of the hypothetical selection process, they are proving invaluable in another area of biology; one that seems to hold out much promise to Darwinists—the field of genetics. Since Watson and Crick elucidated the structure of the DNA molecule, and since geneticists began unraveling the meaning of the genetic code, the center of gravity of evolution theory has gradually shifted away from the earth sciences—geology and paleontology—toward molecular biology.

This shift in emphasis has occurred not only because of the attraction of the new biology as holding the answers to many puzzling questions, but also because the traditional sciences have proved ultimately sterile as a source of decisive evidence. The gaps in the fossil record, the incompleteness of the geological strata, and the ambiguity of the evidence from comparative anatomy ultimately caused Darwinists to give up and look somewhere else for decisive evidence. Thanks to molecular biology and computer science they now have somewhere else to try.

Darwinists seem to have drawn immense comfort from their recent discoveries at the cellular level and beyond, behaving and speaking as though the new discoveries of biology represent a triumphant vindication of their long-held beliefs over the irrational ideas of vitalists. Yet the gulf between what Darwinists claim for molecular biological discoveries and what those discoveries actually show is only too apparent to any objective evaluation.

Consider these remarks by Francis Crick, justly famous as one of the biologists who cracked the genetic code, and equally well known as an ardent supporter of Darwinist evolution. In his 1966 book *Molecules and Men*, in which he set out to criticize vitalism, Crick asked which of the various molecular biological processes are likely to be the seat of the "vital principle."[3] "It can hardly be the action of the enzymes," he says, "because we can easily make this happen in a test tube. Moreover most enzymes act on rather simple organic molecules which we can easily synthesise."

There is one slight difficulty but Crick easily deals with it: "It is true that at the moment nobody has synthesised an actual enzyme chemically, but we can see no difficulty in doing this in principle, and in fact I would predict quite confidently that it will be done within the next five or ten years."

A little later, Crick says of mitochondria (important objects in the cell that also contain DNA):

It may be some time before we could easily synthesise such an object, but eventually we feel that there should be no gross difficulty in putting a mitochondrion together from its component parts.

This reservation aside, it looks as if any system of enzymes could be made to act without invoking any special principles, or without involving material that we could not synthesise in the laboratory.[4]

There is no question that Crick and Watson's decoding of the DNA molecule is a brilliant achievement and one of the high points of twentieth-century science. But this success seems to me to have led many scientists to expect too much as a result.

Crick's early confidence that an enzyme would be produced synthetically within five or ten years has not been borne out and biologists are further than ever from achieving such a synthesis. Indeed, reading and rereading the words above with the benefit of hindsight I cannot help but interpret them as saying "we are unable to synthesize any significant part of a cell at present, but this reservation aside, we are able to synthesize any part of the cell."

Certainly great strides have been made. William Shrive, writing in the *McGraw Hill Encyclopedia of Science and Technology*, says, "The complete amino acid sequence of several enzymes has been determined by chemical methods. By X-ray crystallographic methods it has even been possible to deduce the exact three-dimensional molecular structure of a few enzymes."[5] But despite these advances no one has so far synthesized anything remotely as complex as an enzyme or any other protein molecule.

Such a synthesis was impossible when Crick wrote in 1966 and remains impossible today. It is probably because there is a world of difference between having a neat table that shows the genetic code for all twenty amino acids (alanine = GCA, proline = CCA, and so on) and knowing how to manufacture a protein. These complex molecules do not simply assemble themselves from a mixture of ingredients like a cup of tea. Something else is needed. What the something else is remains conjectural. If it is chemical it has not been discovered; if it is a process it is an unknown process; if it is a "vital principle" it has not yet been recognized. Whatever the something is, it is presently impossible to build a case either for Darwinism or against vitalism out of what we have learned of the cell and the molecules of which it is composed.

It is easy to see why evolutionists should be so excited about cellular discoveries because the mechanisms they have found appear to be very simple. But however simple they may seem, as of yet no one has succeeded in synthesizing any significant original structure from raw materials. We know the code for the building blocks; we don't know the instructions for building a house with them.

Indeed, the discoveries of biochemistry and molecular biology have raised some rather awkward questions for Darwinists, which they have yet to address satisfactorily. For example, the existence of genetically very simple biological entities, such as viruses, seems to support Darwinist ideas about the origin of life. One can imagine all sorts of primitive life forms and organisms coming into existence in the primeval ocean and it seems only natural that one should find entities that are partway between the living and the nonliving—stepping stones to life as it were. It is only to be expected, says Richard Dawkins, that the simplest form of self-replicating object would merely be that part of the DNA program which says only "copy me," which is essentially what a virus is.

The problem here is that viruses lack the ability to replicate unless they inhabit a host cell—a fully functioning cell with its own genetic replication mechanisms. So the first virus must have come *after* the first cell, not before in a satisfyingly Darwinian progression.

But despite minor unresolved problems of this kind Darwinists still have one remaining card to play in support of their theory. It is the strongest card in their hand and the most powerful and decisive evidence in favor of Darwinian evolutionary processes.

PART FOUR

CREATION

CHAPTER 16

Pandora's Box

B Y FAR THE STRONGEST PRIMARY EVIDENCE for evolution, for common descent and for Darwinian processes of mutation and natural
selection, is that of homology. *Homology* is the name given to the
anatomical correspondences between different species that biologists
and paleontologists have noted and studied for centuries.

Darwin himself explained the significance of homology with
eloquent simplicity in *The Origin of Species* when he said,

> We have seen that the members of the same class, indepen-
> dently of their habits of life, resemble each other in the
> general plan of their organisation. This resemblance is of-
> ten expressed by the term "unity of type"; or by saying that
> the several parts and organs in the different species of the
> class are homologous.
>
> What can be more curious than that the hand of a man,
> formed for grasping, that of a mole for digging, the leg of
> the horse, the paddle of the porpoise, and the wing of the
> bat should all be constructed on the same pattern and should
> include similar bones in the same relative position?

On the face of it, there can be only one rational explanation for
such similarities and that is descent from a common ancestor from
whom the similar features are a genetic inheritance. Some homolo-

gies are so striking that it appears impossible to deny this interpretation. Every four-footed vertebrate animal has the same pentadactyl design with the same set of bones in modified form. The bones of the arm, wrist, and hand that are found in humans can also be found in modified form in all other four-limbed animals with backbones.

It is homology that leads Darwinists to put together isolated fossil remains in ancestor-descendant relationships—often very convincing ones. It is homology that Darwinists rely on to bridge the gaps in the fossil record, as in the case of horses. It is homology that underlies the diagrams drawn up by Darwinists from Haeckel to the present day showing how every living thing is related.

Ultimately, however, it is homology that has provided the greatest stumbling block to Darwinian theory, for at the final and most crucial hurdle, homology has fallen.

In the past hundred years, biology has undergone successive revolutions—in embryology, in microbiology, in molecular biology, and in genetics, revolutions which have laid open on the laboratory bench the most minute detail of how plants and animals are constructed. If the Darwinian interpretation of homology is correct, then you would expect to find at the microscopic level the same homologies that are found at the macroscopic level. In fact that is not what has been found.

This fundamental disappointment has been called by Australian molecular biologist Michael Denton "the failure of homology." In his book *Evolution: A Theory in Crisis* Denton says,

> The validity of the evolutionary interpretation of homology would have been greatly strengthened if embryological and genetic research could have shown that homologous structures were specified by homologous genes and followed homologous patterns of embryological development. Such homology would indeed be strongly suggestive of "true relationship; of inheritance from a common ancestor." But it has become clear that the principle cannot be extended in this way. Homologous structures are often specified by non-homologous genetic systems and the concept of homology can seldom be extended back into embryology.[1]

In embryological development, for example, organs that appear identical in different animals do not arise from the same site or group of cells of the embryo. Even a fundamental structure such as the alimentary canal, found in all vertebrates, is formed differently in different animals. In sharks it is formed from the roof of the embryonic gut cavity, whereas in the lamprey it is formed from the floor of the gut; from the roof and floor in frogs; and from the lower layer of the embryonic disc, or blastoderm, in birds and reptiles.[2]

The classic case of homology referred to by Darwin—that of the forelimbs in vertebrates—turns out in fact to be flawed, since forelimbs develop from different body segments in different species. In the newt, the forelimbs develop from trunk segments 2,3,4, and 5; in the lizard from segments 6,7,8, and 9; and in humans from segments 13,14,15,16,17, and 18.[3] As Michael Denton points out, from this evidence it could be argued that the forelimbs are not strictly homologous at all.

Again, according to Denton,

> The development of the vertebrate kidney appears to provide another challenge to the assumption that homologous organs are generated from homologous embryonic tissue. In fish and amphibia the kidney is derived directly from an embryonic organ known as the mesonephros, while in reptiles and mammals the mesonephros degenerates towards the end of embryonic life and plays no role in the formation of the adult kidney, which is formed instead from a discrete spherical mass of mesodermal tissue, the metanephros, which develops quite independently from the mesonephros.

Many other comparable examples can be given from embryology: in almost every case they have been put into a file drawer labeled "unresolved problems of homology" and largely forgotten about.[4]

It isn't only embryology that experienced such disappointments. In the 1950s, when molecular biologists began to decipher the genetic code, there was a single glittering prize enticing them. When they found the codes for making proteins out of amino acids, they naturally assumed that they were on the brink of discovering at the molecular level the same homologies that had been observed at the macroscopic level in comparative anatomy.

If the bones of the human arm could be traced to the wing of the bat and hoof of the horse, then the miraculous new science of molecular biology would trace the homologies in DNA codes that expressed these physical characteristics. At long last, biologists were on the brink of opening Pandora's box and finding inside the final key to life: the chemical formula for an arm or a leg or an eye.

Yet when biologists did begin to acquire an understanding of the molecular mechanism of genetics, they found that apparently homologous structures in different species are specified by quite different genes. Pandora's box turned out to be empty.

The main problem with understanding the genetic code contained in the DNA molecule is that individual genes do not appear to correspond to individual characteristics. The gene that controls the color of a mouse's coat also controls the mouse's size. The gene that controls the color of the eye of the fruit fly *Drosophila* also controls the shape of the female sex organs. Almost all genes in higher organisms have multiple effects of this sort and Ernst Mayr has suggested that genes which control only a single characteristic must be rare or nonexistent.[5]

Denton gives an example of the multiple effects of a single gene in the case of the domestic chicken. There is a degenerative mutation known for a single gene that causes a wide range of defects: no proper development of the wings; no claws on the feet; underdeveloped covering of downy feathers; lungs and air sac absent. The significance of this case is that some features affected are unique to birds (wings, feathers) while others, such as the lungs, occur in many other vertebrate species including humans.

Denton points out that "this can only mean that nonhomologous genes are involved to some extent in the specifications of homologous structures."

There are other puzzles contained within homology both in principle and in practice. For instance, humans—and other four-limbed vertebrates—have hind limbs which are exactly homologous in structure to their forelimbs. Yet this cannot possibly be taken as evidence of common descent. Human hind limbs cannot have descended from human forelimbs. Moreover, if vertebrate limbs have evolved from fish anatomy then they must have evolved from different precursors: the forelimbs from the pectoral fins of the fish, the hind limbs from the pelvic fins. Yet today they are identical, apparently homologous, structures.

The only explanation that Darwinists have to offer is that both forelimbs and hind limbs represent a case of "convergent" evolution, although, once again, this is not so much an explanation as an example of tautology being pressed into service to cover a gap in our knowledge.

The remarkable discoveries of biochemistry and molecular biology since the 1950s have provided much evidence that, on first reading, appeared to support many of the premises of Darwinism. For example, there are some proteins that are widely used in many organisms, such as the proteins cytochrome C and hemoglobin. Research showed that the sequences of amino acids comprising these proteins varied slightly from species to species. This seemed enormously promising for it appeared to show a variation at the molecular level between species that would mirror the morphological differences in the anatomy of those species. Although fossils and comparative anatomy had failed, biochemistry could perhaps provide the evidence Darwinists sought of patterns of evolutionary inheritance.

It was discovered, for example, that the similarity between the hemoglobin sequences of animals thought by Darwinists to be more closely related was greater than that of creatures thought to be distantly related. This confirmed the Darwinian view of genetic relationships. When the hemoglobin sequence of two mammals such as a human and dog were examined, they were found to have a divergence of only about 20 percent, whereas when the hemoglobin of a human and a fish were examined, they were found to diverge by more than 50 percent.

Perhaps by compiling a table of sequences of all the common proteins for all species we could get a quantified numerical picture of how closely or distantly related each species is?

This hope, too, was dashed. According to Michael Denton,

As more protein sequences began to accumulate during the 1960s, it became increasingly apparent that the molecules were not going to provide any evidence of sequential arrangements in nature, but were rather going to reaffirm the traditional view that the system of nature conforms fundamentally to a highly ordered hierarchic scheme from which all direct evidence for evolution is emphatically absent.

What biochemists found when they compiled their table of proteins (such as cytochrome C) is that it is possible to classify species into groups and that these groups do indeed correspond exactly to the groups that have been arrived at by comparative anatomy. However, what is most striking about such a protein "atlas" is that each of these identifiable groups or subclasses is isolated and distinct from the others. There are no transitional or intermediate classes, just as there are no transitional species in the fossil record or in the living world today.

Denton points out that published tables of divergence of the cytochromes, such as the *Dayhoff Atlas of Protein Structure and Function*, illustrate this dramatic absence of intermediates.[6]

The most primitive organisms are bacteria whose cells do not contain a nucleus. All higher organisms, from yeasts to humans, whose cells do contain a nucleus, are called eukaryotes. If all eukaryotes have descended from bacteria, then you would expect to find a graduated divergence in their proteins like cytochrome C. In fact what you find is that all the main classes, from man to kangaroo, from fruit fly to chicken, from sunflower to rattlesnake, and from penguin to baker's yeast, are all equidistant from bacteria with around 65 to 69 percent divergence.

According to Denton,

Eucaryotic cytochromes, from organisms as diverse as man, lamprey, fruit fly, wheat and yeast, all exhibit a sequence divergence of between sixty-four per cent and sixty-seven per cent from this particular bacterial cytochrome. Considering the enormous variation of eucaryotic species from unicellular organisms like yeasts to multicellular organisms such as mammals, and considering that eucaryotic cytochromes vary among themselves by up to about forty-five per cent, this must be considered one of the most astonishing findings of modern science.[7]

Even more extraordinary is the complete absence of evidence from biochemistry for the most basic Darwinian evolutionary scheme of fish to amphibian to reptile to mammal. When the protein divergence of land-dwelling vertebrates—amphibians, reptiles, mammals—are compared with those of fish, they are all again equally isolated. There is no graduation of divergence as one would expect in an evolutionary sequence.

The horse, the rabbit, the frog, and the turtle are all 13 percent divergent in their cytochrome C from the carp. "At a molecular level," says Denton, "there is no trace of the evolutionary transition from fish to amphibian to reptile to mammal. So amphibia, always traditionally considered intermediate between fish and the other terrestrial vertebrates, are in molecular terms as far from fish as any group of reptiles or mammals."

Perhaps the most baffling finding of all is that radically different genetic coding can give rise to animals that outwardly look very similar and exhibit similar behavior, while creatures that look and behave completely differently can have far less genetic divergence. There are, for instance, more than 800 species of frogs, all of which look superficially the same. But there is a greater variation of molecular structure between them than there is between the bat and the blue whale.[8]

Denton points out that perhaps the greatest irony regarding modern discoveries in molecular biology is that, had this information been available a century ago to opponents of Darwin such as Richard Owen and Louis Agassiz, then Darwin's ideas on evolution might very well never have been accepted at all.

CHAPTER 17

Paradigm Lost

IN 1962, THOMAS KUHN ASTONISHED his academic contemporaries by proposing that scientific theories should be looked on
not only as dealing with pure objective facts, but rather as systems
of belief relating to a wider context: a frame of reference consisting
of interlocking scientific, social, and even political ideas. This ideological context, which Kuhn terms a paradigm, is implicitly agreed
upon by scientists who subscribe to a particular theory and who
share the same world view.

The power of such a paradigm, says Kuhn, is so great that some
scientists will continue to believe it even in the face of contradictory evidence (a phenomenon dubbed cognitive dissonance by psychologist Leon Festinger). This blinkered dogmatism continues
until new evidence is overwhelming and a new theory deposes the
old—a "global paradigm shift" occurs.

Such an ideological context can be found in anthropology in the
nineteenth century when most Victorian scientists shared the implicit belief that the colored races were genetically inferior to the
white European race. Because the belief in the genetic inferiority of,
for instance, the Australian aborigine was widely shared by scientists,
then scientific "evidence" was brought forward to substantiate this
viewpoint and was generally accepted. Textbooks illustrated evidence
of the aborigine's Stone-Age level of cultural attainment, his coarse
features, supposed low intelligence, and brutal behavior.

185

Thomas Huxley, who was Darwin's leading supporter, observed that "No rational man, cognizant of the facts, believes that the Negro is the equal, still less the superior, of the white man."

Darwin himself founded much of his evolutionary thinking on equally racist ideas. In *The Descent of Man* he indicated his belief that the Negro races were more closely related to the apes than white people and also his belief that, "at some future period, not very distant as measured by centuries, the civilised races of man will almost certainly exterminate and replace the savage races throughout the world."

Today, few scientists would maintain that such beliefs were justifiable on grounds of observation and measurement—not because the evidence has changed, but because the ruling paradigm of anthropological science has changed. Institutionalized racism withered under the twin effects of the decline of imperialism and the rise of civil rights movements.

One consequence of a scientific theory occurring in an ideological context is that much of the evidence which apparently supports that theory actually merely supports its acceptability to scientists and members of the community. Few people in Darwin's day questioned the belief in the inferiority of the Australian aborigine, simply because that belief was part and parcel of the world view of Europeans of the imperial Victorian age. But that implicit belief became in turn the foundation for the *scientific* view that all the races of mankind represented an evolutionary spectrum, ranging from the genetically "undeveloped" aboriginal type to the genetically "advanced" white European type. This appeared to be evidence in favor of the theory of Darwinian evolution itself.

In much the same manner, Darwin's theory became buttressed at an early stage by a powerful array of supporting evidence, held to confirm its basic principles, but which in fact represented nothing more than the assumptions of the ruling ideology of Darwin's era. These assumptions concerned a broad range of minutely described natural phenomena, such as the persistence of vestigial organs in the human body, left behind by evolution, and the recapitulation of former evolutionary stages by embryos.

Since the ruling ideology, the paradigm, of the life sciences has changed since Darwin's day, the assumptions and ideas which formerly acted as supporting evidence for his theory have melted away like snow on a spring morning. The large mass of peripheral evidence for the theory has been gradually eroded by further discoveries, more accurate observation, and science's changing world view. As is often the case with Darwinism, however, although these former assumptions have been exposed as without foundation, they somehow remain in the popular evolution mythology and continue to be referred to in textbooks and lectures.

A case in point is the existence of "vestigial" organs in the human body: organs deemed by evolutionists to have become redundant through the action of evolution. In his 1895 book *The Structure of Man*, Ernst Wiedersheim lists eighty-six organs of the human body which were supposed to have lost their function, and to be mere appendages which time and further evolution would no doubt dispel entirely from the human frame. (*Encyclopaedia Britannica* currently gives the total of redundant human organs as "more than 100.") The list includes organs such as the pineal gland, the thyroid gland, the thymus, the coccyx, the appendix, the ear muscles and the tonsils.

The claims for vestigial organs have been examined by S. R. Scadding of the department of zoology at the University of Guelph, Ontario. Scadding's principal conclusion is that, "on the basis of this analysis, I would suggest that Wiedersheim was largely in error in compiling his long list of vestigial organs. Most of them do have at least a minor function at some point in life."[1]

A prime example is the pineal gland, located between the hemispheres of the brain and long believed to be a degenerate eye serving no function. Although still something of a mystery, the pineal body is now known to be an endocrine gland (one that works through the bloodstream) and is thought to be of importance in triggering growth cycles and sexual development in a young individual. It is currently thought that the pineal gland secretes a hormone called melatonin and that this in turn regulates sexual development.

Similarly, the functions of the thyroid gland and thymus were previously unknown and they were assigned "vestigial" status, until their true functions were elaborated. Removal of the thymus gland in adults has no effect and so it was considered without function. In the 1970s it was discovered that the thymus makes a vital contribution in early infancy to the development of the body's immune system. The thyroid, too, is now known to be an endocrine gland which secretes two hormones vital to metabolism and growth.

Of two famous examples—the appendix and the coccyx—Scadding says,

> Anatomically the appendix shows evidence of a lymphoid function since the submucosa is thickened and almost entirely occupied by lymphatic nodules and lymphocytes. There is experimental evidence as well that the vermiform appendix is a lymphoid organ which acts as a reservoir of antibody producing cells. The coccyx serves as a point of insertion for several muscles and ligaments including the *gluteus maximus*. Similarly, for other "vestigial organs" there are reasonable grounds for supposing that they are functional albeit in a minor way.

From the point of view of human anatomy studies, it matters little that an organ is believed to be useless but is later discovered to have a useful function. From the point of view of evolution theory, it matters considerably, since the supposed "vestigial" character of such organs has been adduced as evidence of evolution in action. It remains to be seen how many other human organs which are currently supposed to be vestigial, will turn out to have equally important functions. In the mean time, it would be unscientific, to say the least, to claim them as vestigial.

Once again, few scientists today would take seriously such arguments. But as usual, the existence of vestigial organs is still referred to in school biology lessons and some textbooks, because it seems reasonable. Simpson, for example, in his book on the evolution of horses, describes the human coccyx as being a vestigial organ, homologous with the ape's tail and with no modern purpose.[2]

Like other branches of science, Darwinism has been led down some seriously wrong roads over the past century by overenthusiastic individuals. No errant scientist has been more thoroughly disowned by his colleagues than German zoologist Ernst Haeckel. Haeckel performed a service to zoology by coining the handy term "ecology." Unfortunately he also conceived the "biogenetic law"—

the idea that the developing embryo passes through or recapitu-lates stages in the evolution of its entire phylum (its ancestral tribe or race). An unstoppable creator of neologisms, Haeckel asserted in his 1876 book *General Morphology of Organisms* that "ontogeny recapitulates phylogeny."

What Haeckel (and a substantial number of early followers in-cluding Darwin) believed was that the human embryo started life resembling a single-celled marine organism, then developed into a worm with a pulsating-tube heart, then into a fish with gill slits and a two-chambered heart, then into an amphibian with a three-chambered heart, into a mammal with a four-chambered heart and a tail (for swinging through the trees), and finally into a human baby. These various stages involved the embryo exhibiting vesti-gial remnants of former evolutionary stages (such as gills) which it was obliged by natural law to pass through in order to reach its new, higher stage of evolution.

The biogenetic law is no longer taken seriously by embryolo-gists. But once again the idea has passed into evolutionary myth and is still to be found in some textbooks and is also referred to in school and university lectures. Although abandoned as having the status of a scientific law, the feeling persists that there is "some-thing in it." The trouble with Haeckel's law is that the observations it seeks to offer as evidence for evolution theory come not from nature but from a human viewpoint, and rely for their force on purely superficial resemblances. The human embryo is never a single-celled marine organism, nor does it ever live in a marine aquatic environment. It never possesses gill-slits nor does it ever breathe, rather it takes oxygen directly from its mother's blood stream. The human embryo does develop folds of skin superficially resembling gills but they are not gills. They are structures that become the lower jaw, tongue, and other organs of the throat.

Equally, the order in which events occur superficially re-sembles the order of supposed evolution but, as Dr. A. J. White has pointed out, is actually different.[3] It is true that the human embryo begins with a single-chambered heart which develops into two chambers. But this early structure then reverts to a single chamber again before redeveloping later to two, three, and fi-nally four chambers. No Darwinist has so far suggested that the phylum from which humans descended underwent evolution from a two-chambered heart to a single chamber, since this would be a backward step and bad news for Darwinism.

Although no professional scientist today would consciously admit to believing in the biogenetic law, even the most eminent Darwinists are prone to slip into this way of thinking when not on their guard. Gavin de Beer was a professional embryologist (as well as Director of the British Museum of Natural History). In his 1964 *Atlas of Evolution* he is careful to disavow Haeckel's "law" that ontogeny recapitulates phylogeny. But in the same book, while describing the evolution of the eye, he remarks, "There can be little doubt that the series of stages . . . through which the eye passes in embryonic development is a repetition of the manner in which it evolved."[4]

The concept of recapitulation of past evolutionary stages was an important one for evolution theory not only in embryology, since it could be used, and was used, to explain a wide range of common observations from the natural world which contradict the fundamental idea of progressive evolution. If species become progressively better adapted to their environment (by developing eyes with lenses and color vision for instance, or by developing a shape and coloring which mimics other creatures or other natural objects) then one would expect the fossil record to show such cumulative complexity through time.

In fact this is not what the fossil record shows. Sometimes the anatomy of creatures becomes more complicated (often in bizarre and apparently senseless ways like the skulls of some dinosaurs or the gigantic antlers of the Irish elk) but they are succeeded in the rocks by remains of creatures who become simpler again and then complicated again. The solution to this mystery, said Darwinists, was that the simpler creatures were merely degenerate recapitulations of their ancestral forms. As described in an earlier chapter, Hyatt arranged the ammonite sequence of the Liassic rocks in a palpably incorrect order, based solely on the concept of recapitulation.

"Recapitulation" was useful to evolutionists in other ways. If it were possible, even in principle, for an organism to revisit anatomical characteristics of its phylum, then perhaps we might seek clues to each creature's descent through these anatomical throwbacks? Thus some evolutionists have postulated such phenomena as "proterogenesis" (the appearance of ancestral features in the young of the species) and "pedomorphism" (anatomical features in modern species resembling the young stage of an ancestral species—a kind of Peter Pan syndrome).

Some Darwinists have been profoundly disenchanted by this proliferation of so-called evolutionary effects and have even been moved to complain. An obviously exasperated Mayr wrote in 1960, "the attempt to 'explain' genetic and selective processes by all sorts of fancy terms like 'pedogenesis,' 'palingenesis,' 'proterogenesis,' and whatnot have had a stultifying effect on the analysis. The less said about this type of literature, the better."[5]

Most of these false avenues are no more than minor embarrassments to Darwinists. However, it is the concept of "convergence" (considerably more important than those complained of by Mayr) that highlights one of the greatest weaknesses in the synthetic theory and one which, though raised before, has yet to be satisfactorily addressed—unless it is to become a suitable candidate for Kuhn's global paradigm shift. The weakness has to do with the geological events referred to in an earlier chapter, the breakup of the original supercontinent of Pangaea into the present-day land masses, thus separating the plant and animal populations of those continents and—according to Darwinists—allowing them to evolve in isolation. Uniformitarians place this event toward the close of the Mesozoic era, that is somewhere in the region of 65 million years ago, according to currently accepted geochronometry.

At the time the present continents were formed, the life they contained was very different from life today. The dominant life forms were the dinosaurs. The only representatives of the mammals (our own branch of the animal kingdom) then alive were tiny shrewlike creatures. It has been proposed that the reptiles dominated every available ecological niche so effectively that the mammals were hardly able to get a toehold (Harvard's Stephen Jay Gould describes them as living in the nooks and crannies of the reptilian world). It was only after the mysterious mass extinction of dinosaurs and thousands of other species at the end of the Mesozoic era that mammals were able to begin their rise to dominance, culminating in the appearance of humans.

Practically all the mammals that have appeared are either placental (bearing young until fully developed, like humans) or marsupial (giving birth prematurely and nurturing the young in a pouch, like kangaroos). The marsupial mammals are confined to Australia and South America, and are said to have evolved uniquely in those environments, while at the same time placental mammals were evolving elsewhere.

The key factor about the evolution of the marsupials is that a large number of modern marsupial animals exist which—apart from the pouch and child-rearing habits—are identical with placental mammals to an extraordinary degree. This is no mere general similarity of anatomical detail, but an almost perfect duplication of distinctive species like cats, rats, wolves, moles, flying squirrels, anteaters, and others. In addition there are distinctive marsupials which exist only in Australia, such as the koala and the kangaroo.

How does it come about that in widely separated environments the same tiny shrewlike ancestral mammal of 65 million years ago should evolve on strictly parallel lines to produce virtually the same range of large mammals today? The Tasmanian marsupial wolf is a virtual carbon copy of the European timber wolf. The marsupial flying phalanger is practically identical to the placental flying squirrel, as are the marsupial jerboa and the placental jerboa. When the skulls of the two wolves are placed side by side, it would take an experienced professional zoologist to tell them apart.

The question for Darwinists is: How can a mouselike creature have evolved into two identical wolflike creatures (*and* two identical moles, etc.) on two different continents? Doesn't this coincidence demand not merely highly improbable random mutations, but miraculous ones? According to Simpson in *The Meaning of Evolution* the answer is simple. This convergence comes about through the "selection of random mutations."[6]

In a different book, the same author concludes that "Tasmanian and true wolves are both running predators, preying on other animals of about the same size and habits. Adaptive similarity involves similarity also of structure and function. The mechanism of such evolution is natural selection."[7]

As Arthur Koestler observed of this example, "One might as well say, with the wisdom of hindsight, that there is only one way

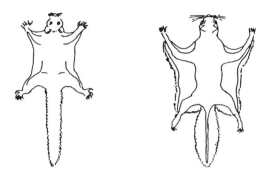

Marsupial flying phalanger (left) and placental flying squirrel.

Marsupial jerboa (left) and placental jerboa.

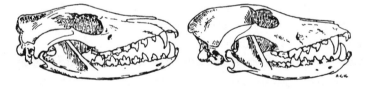

Skulls of marsupial Tasmanian wolf (left) and placental wolf.

The Australasian marsupials on the left are among many species identical with their placental counterparts. Darwinists believe these species have evolved in parallel but quite separately over the last 65 million years from a common shrew-like ancestor. By random mutation? *(From* The Living Stream *by Sir Alister Hardy)* of making a wolf, which is to make it look like a wolf."[8]

The stupendous inadequacy of Simpson's explanation, and the almost casual way in which Darwinists have batted aside the marsupial problem is, I think, a symptom of their uneasiness over the issue—perhaps a symptom of cognitive dissonance. The response reveals the synthetic theory's inability to explain a key real-life biological problem. But more than this, the existence of identical evolutionary outcomes in isolated environments is the strongest possible indication that random mutation and natural selection are incapable of explaining the origin of species. At the same time these outcomes are the strongest possible indicator of some other important process or processes at work, which somehow limit or direct the repertoire of evolution. An exploration of what those processes may be will have to wait until a later chapter.

As far as most people are concerned, of course, arguments concerning wolves and jerboas are of only academic interest. The big issue as far as people are concerned is the origin of humans. One might expect that the overwhelming majority of the efforts of evolutionary biologists today would be devoted to this subject. Yet, paradoxically, we ourselves have become almost a neglected subject.

The more exciting and more promising world of the microscope and the computer, coupled with discoveries in molecular biology, have meant that microbes and molecules have replaced humankind as the focus of the evolutionists' attention.

In contrast to the first decades of the twentieth century, when human anthropology and the study of human evolution was the most important subject of scientific study, it has today been relegated to virtual obscurity and has become the province of a relatively small number of talented individuals, working in isolation.

In some respects, the evolution of the human species has become almost a taboo subject: too hot to handle politically, and equally dangerous scientifically. Riddled with doubt and smarting from numerous embarrassing mistakes and forgeries, like Piltdown man, evolutionists have quit the field almost entirely. And apart from heroic individual efforts like the Leakeys and Johanson working in Africa, there have been few significant paleontological efforts relating to humans since the Second World War.

You might imagine that the effect of this disillusion and abandonment would be reflected in schools, universities, and museums by a reduced interest in human evolution and a dearth of teaching materials. Yet the old myths are more active than ever, and reconstructions of our apelike brutish ancestors and their primitive lives form part of schoolboy folklore in the 1990s, just as they did in the 1930s. Even in major works of science fiction, like Arthur C. Clarke's popular *2001: A Space Odyssey*, or Pierre Boule's *Planet of the Apes*, it is taken for granted that our ancestors were apelike. And the writers and illustrators of popular works of historical geology spend considerable time and effort searching for minute "accuracy" in their reconstruction of apelike paleolithic hunters and their environment.

Strangely too, this modern confidence and apparent precision in reconstruction is not based on further discoveries of fact, but takes place *despite* the discoveries of recent decades—that the evidence for humankind's own evolution is actually nonexistent.

CHAPTER 18

Down from the Trees

Less than a decade after the publication of the *Origin of Species* in 1868, Ernst Haeckel published a monumental work, primarily inspired by Darwin, and grandly entitled *The History of Creation* in which he fully indulged his predilection for coining new words. This time, Haeckel became even more ambitious and coined not merely a new scientific term but a new generic and specific name for a living creature, but a creature no one had ever seen and for which there was no evidence at all—*Pithecanthropus alalus*, the "speechless ape-man." *Pithecanthropus*, said Haeckel, was the link between humans and our apelike ancestors. When his or her fossil remains were discovered, they would be found to have some ape-like characters and some human characters.

Haeckel was even able to describe some of these characteristics: long arms, short legs with knock knees, a half-erect walk, and a long skull with slanting teeth. And the great biologist was also able to suggest the area of the world where the remains were most likely to be found: the hypothetical ancient continent of Lemuria, stretching from Madagascar to India and across the Indian Ocean to Indonesia. However, added Haeckel gloomily, it is "ridiculous to expect paleontology to furnish an unbroken series of positive data."

Haeckel's pessimism on this point was proved unjustified by events. For within decades of his prediction, *Pithecanthropus* was found, possessing just the characteristics Haeckel had predicted, and in the very spot he had foretold.

An ambitious and talented Dutch anatomist, Eugene Dubois, set sail with his wife and young children in 1887 for the Dutch colony of Java in the East Indies. Dubois had signed up for a spell of service with the Dutch army medical corps. A teacher of anatomy, Dubois was thoroughly versed in the works of Haeckel, and as an experienced geologist and paleontologist he was perfectly equipped for the task on which he embarked. The young man was setting out with the avowed intention of being the first to discover concrete evidence of the missing link between humans and apes; the first to discover *Pithecanthropus*.

Within two years of arriving in Sumatra, Dubois had persuaded the government to allow him to carry out a complete paleontological survey of Java under his full-time supervision. He was given convict workers to carry out excavations and military personnel to supervise the digging. Dubois himself, however, did not participate in the fieldwork and contented himself with examining each season's finds on the veranda of his house where they were periodically delivered by the convict crew.

In 1891 Dubois made two important finds amongst the bones dumped on his veranda. The fossils were a tooth and a skullcap which had been found a month apart in the same fossiliferous bed but in locations that were not known exactly because no one was recording the finds. At first, Dubois identified them as belonging to a chimpanzee. Some months later, however, the convicts found a fossil thighbone in the same bed, a thighbone which belonged unmistakably to an upright walking human. Dubois now revised his earlier identification and put the femur together with the skullcap and tooth to produce *Pithecanthropus erectus*—"upright apeman"—a vindication of Haeckel and the first solid evidence of a missing link. Great efforts were made to secure further finds. Some 10,000 cubic meters of sediments were dug and sieved, but the only additional discovery was another tooth.

The third International Congress of Zoology at Leiden in 1895 greeted the fossils with unanimous recognition as to their importance, but with a mixed reception regarding their interpretation. The president, Rudolph Virchow (founder of the modern science

of pathology), cast doubt on the remains belonging to a single individual. Some members felt they were more ape than human, while
others felt they were entirely human. A few agreed with Dubois
that he had bagged a missing link. Haeckel, who was present, was
delighted to have been proved right but was rather circumspect
about the finds: "Unfortunately, the fossil remains of the creature
are very scanty: the skullcap, a femur, and two teeth. It is obviously
impossible to form from these scanty remains a complete and satisfactory reconstruction of this remarkable Pliocene Primate."[1]

If he was publicly cautious, though, Haeckel was privately convinced because he paid from his own pocket for a life-size reconstruction to be built which stands today in the basement of the
Leiden Natural History Museum. In common with all reconstructions of ape-men, both three dimensional and pictorial, which have
been essayed since Haeckel and Dubois's day, the Leiden statue
bears a humanlike body and a rather dim, apelike face. He is gazing
with a puzzled frown at a crude knifelike tool clutched in his primitive hand, as though trying to remember with his small brain how
he came to be relegated to the museum basement from the fashionable salons upstairs.

Though his statue has been edged away from the public gaze,
Haeckel's *Pithecanthropus* and the creature's familiar epithet "Java
man" still figure prominently in evolutionary mythology, a testa-
ment to the staying power of a good story, whatever its true merits.
Today, "Java man" is thought to be an extinct, giant gibbonlike
creature and not connected to humans.

The story of Dubois's discovery of Java man, like Gideon
Mantell's discovery of the first dinosaur, is a parable of primate
paleontology in the past 100 years. Discoveries are few and fortuitous, yet it is extraordinary how they are always deliberately sought
by their discoverers. The reconstructions, the names bestowed, and
the attributions to human or ape inheritance blow this way and
that in the wind of scientific opinion. In the end each find has its
supporters and detractors but settles nothing.

This question of attribution has bedeviled every "missing link"
discovery of the twentieth century. The pattern is a recurring one.
The remains themselves are always meager. The first attribution is
always that the being whose remains have been discovered shows

both human and ape characteristics, and is therefore a genuine transitional type—a real missing link. Then the attribution is questioned: the characters ascribed to apes are actually within the range of human characters; or ape remains postdate the finds by a large margin; or the reconstruction work is over imaginative; sometimes simple mistakes of identification are made perhaps due to disease or malformation of bones.

The position today is that all the fossil remains which were previously assigned some intermediate status between apes and humans have later been definitely reassigned into the categories of either extinct ape or human, and this reassignment has been accepted by all but the most fanatical devotees of this or that fossil.

Strangely enough, although evolutionists from Darwin onward have frequently harped on how unfair it is that vertebrate remains are very rare and their discovery a matter of chance, the world's natural history museums are today bulging with vertebrate remains from Europe, Asia, and Africa. Yet as with all other branches of the animal kingdom, the gaps remain where there should be transitional species. In the case of humans, there is not just one gap to be filled (between apes and ourselves) but many gaps.

First there is the gap between mammals and the rest of the animal kingdom. So far there is no candidate for the ancestor of all the mammals except a hypothetical one. No fossil remains have been found. As A. J. White points out, there are a number of distinctive anatomical differences between reptiles and mammals, chiefly the articulation of the jaw (mammals have a single lower jawbone while reptiles have six) and the mechanism of the ear (mammals have three ear bones, reptiles have one). So recognizing a transitional skeleton ought to be straightforward if, as Darwinists claim, mammals evolved from reptiles.[2] The earliest mammals are small rodentlike animals, but there is no evidence in the fossil record for the evolution of rodents or any other group of mammals. According to A. S. Romer of the University of Chicago, in his text book *Vertebrate Paleontology,*

> The origin of the rodents is obscure. When they first appear, in the late Paleocene, in the genus *Paramys,* we are already dealing with a typical if rather primitive true rodent

Restorations of "Piltdown man" (top), Java man (middle), and Neanderthal man (bottom). Darwinist restorations based on fragmentary finds of bones and teeth always manage to convey a distinct "missing link" quality to their former owners. The convincing Piltdown man is wrongly based on a simple forgery associating a normal human skull with the jaw of an ape. But the same "artistic skill" and imagination have been applied to the genuine fossils. (*Restorations by J. H. McGregor, from* Men of the Old Stone Age *by Henry Fairfield Osborn*)

with the definitive ordinal characters well developed. Presumably of course they had arisen from some basal, insectivorous, placental stock, but no transitional forms are known.

The same is true of all other mammals from bears to whales and from walruses to carpenters.

Next there is the gap between the primates and the rest of the mammals. Again, the candidate for this honor is a hypothetical insectivore but no remains of this ant-eating ancestor have ever been found. According to A. J. Kelso in his *Physical Anthropology*, "The transition from insectivore to primate is not documented by fossils. The basis of knowledge about the transition is by inference from living forms."[3]

Then there is the crucial gap: the gap between the hypothetical apelike primate ancestor and ourselves. Despite scores of candidates, the glass cabinet marked "missing link" remains tantalizingly empty. No primate paleontologist has gone on record as admitting such a heretical thought but it is hard to resist the conclusion that it is now likely to remain empty. Richard Leakey has quoted fellow paleontologist David Pilbeam as saying,

> If you brought in a smart scientist from another discipline and showed him the meagre evidence we've got he'd surely say, "forget it; there isn't enough to go on." Neither David nor others involved in the search for mankind can take this advice, of course, but we remain fully aware of the dangers of drawing conclusions from evidence that is so incomplete.[4]

To illustrate the dangers of drawing such conclusions, here is a summary of the stories of just a few fossils who had their fifteen minutes of fame in the glass case before being relegated, like Java man, to the basement of history.

Probably the most celebrated supposed ancestor of modern humans is the unfortunate gentleman whose remains were discovered by quarrymen in a gravel pit in the Neander Valley, near Dusseldorf in 1875. The skullcap and limb bones of "Neanderthal man" were launched on the world by Hermann Schaffhausen, professor of anatomy at Bonn University, who simultaneously introduced into the language a new synonym for coarse, unintelligent brutality.

Neanderthal man was depicted as a shambling brute, who walked with an apelike gait, on the edge of his feet, his low-sloping brow denoting his retarded mentality and antisocial tendencies. He was unquestionably, thought Schaffhausen, part-ape, part-human, and ancestor to modern humans.

It was not until the 1950s, by which time many similar remains had been found in Europe, Africa, and Asia, that Neanderthal man was seriously reevaluated. It was found that some of the original type material belonged to an individual whose bones were thickened and deformed by osteoarthritis and that Neanderthal man's posture was probably the same as modern humans. Evidence was also found that, far from predating Cro-Magnon (modern) humans, the Neanderthals lived at the same time and possibly mixed freely with Cro-Magnons.

Neanderthals sewed clothes from animal skins, used fire for cooking, built shelters, and gave their dead a ritual interment which included placing flowers in the grave. Finally it was observed by Cave and Strauss writing in *The Quarterly Review of Biology* that if he were given a bath, a collar, and tie he would pass unnoticed in the New York subway. Today Neanderthal man is classified as a member of the species *Homo sapiens* and any of us could be among his descendants.

Raymond Dart was a young Australian anatomist who was appointed professor at Witwatersrand University, Johannesburg, in 1922. Dart's speciality was the evolution of the brain and nervous system and he had worked under Grafton Elliott Smith at University College London on developing the technique of making endocranial casts (casts of the inside of skulls) to get an indication of the development of the brain within. Coincidentally, this skull skill was to play a prominent part in his discovery.

In South Africa Dart arranged for workers at the nearby Taung quarry to send him any fossils they found in the limestone rocks being quarried for building stone. In one batch of fossils shipped to his office, Dart immediately recognized a natural endocranial cast made by limestone filling a fossil skull, together with some fragments of the skull itself. It has been remarked before that this endocranial cast had fallen into the hands of one of the three or four people in the entire world capable of recognizing its significance.

Dart believed the cast to show distinctly hominid features in the brain structure, far in advance of any living ape, yet still small and underdeveloped for a human. Dart felt that, "by the sheerest good luck," he had been given "the opportunity to study what would probably be the ultimate answer in the study of the evolution of man."

He wrote a paper for *Nature* and christened his discovery *Australopithecus africanus*—southern ape-man. (Dart was here following the precedent of the American Museum of Natural History who, as we saw in chapter 15, christened a pig's tooth *Hesperopithecus*—western ape-man). Ironically, Dart's discovery was scorned by his scientific contemporaries, partly because of his irreverent Aussie style, but mainly because his identification rested on specialized knowledge of endocranial casts that only he and two or three others possessed.

Dart's discovery was taken up later and championed by Robert Broom, who discovered many more *Australopithecus* remains, which Broom believed showed among other things that the creature walked upright.

Today, despite a century of "missing link" newspaper headlines, *Australopithecus* is the only fossil find which stands any chance at all of being placed in the missing link category and is enthusiastically described by many Darwinists as ancestral to humans.

The real status of *Australopithecus* as an extinct ape was established as long ago as 1954 by the comparative anatomy research of zoologist Solly Zuckermann and his colleagues. Zuckermann compared in detail three diagnostic characteristics in the bones and teeth of *Australopithecus*, in modern humans and in various apes including the gorilla, chimpanzee, orangutan, gibbon, and others. The key characteristics are the size of the brain; the jaws and teeth; and the posture of the head.

By measuring the skulls and teeth of a large number of apes, of fossil *Australopithecines*, and various human specimens, Zuckermann found that *Australopithecus's* head was balanced like that of an ape, not a human; its brain was the same size as modern apes such as the gorilla; and its jaws and teeth are predominantly apelike.

According to Zuckermann,

In the first place, our safest inference from the available facts is that the brains of the fossil *Australopithecinae* did not differ in size or conformation from those of such modern apes as the gorilla. In the second we may conclude that the fossils provide no significant evidence of the major decrease in size of jaws and teeth which is presupposed by the thesis that the *Hominidae* evolved from non-human primate forms. And thirdly the evidence is also clear that the skull of the *Australopithecinae* was balanced on the vertebral column as in apes rather than as in man.

Zuckermann's conclusion is that

The safest overall inference that can be drawn from the facts which have been discussed here is that the *Australopithecinae* were predominantly ape-like, and not man-like creatures.[5]

Identical conclusions were reached more recently by Dr. Charles Oxnard, professor of anatomy and human biology at the University of Western Australia, who in 1984 conducted a computer analysis of Australopithecine fossils. Oxnard, who is a Darwinist, concluded in his 1984 book, *The Order of Man*, that *Australopithecus* is an extinct ape and is unconnected with humankind's ancestry.[6]

Shortly after Zuckermann's study was published, *Australopithecus* was eclipsed from the headlines, not because of Zuckermann's scientific findings but because of excitement over more so-called missing links—this time from East Africa, where Louis Leakey, his wife Mary, and his son Richard have made many discoveries in the region around Olduvai Gorge. The principal cause of the excitement was that the Leakeys' discoveries were made in volcanic deposits which, unlike the sedimentary limestones of South Africa, could be dated by the newly developed potassium-argon method, and using this method yielded a date for the Olduvai Gorge finds of no less than 1.75 million years. This was news indeed. At just the same time, in 1959, Mary Leakey found at Olduvai an almost complete skull which her husband announced to the world as *Zinjanthropus*, East African man.

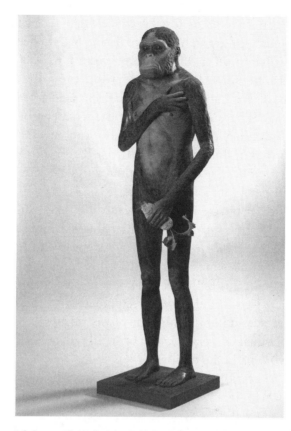

Restoration of *Australopithecus,* a relative of "Lucy," in London's Natural History Museum. *Australopithecus* is shown with humanlike hands and feet, yet the real fossils show Lucy and her relatives to be tree-dwelling apes with hands and feet even longer and more curved than a chimpanzee's, for grasping branches. Artistic license? Or Darwinian myth? *(Photo: Natural History Museum, London)*

The new formula name, with its "anthropus" ending was chosen by Leakey who insisted that his discovery was entirely novel, was not related to Dart's *Australopithecine* discoveries in the South, and was definitely hominid, not an ape. Alas, *Zinjanthropus,* too, fell victim to the curse of all missing links. In 1965 Professor Philip Tobias of Witwatersrand University examined, measured, and described the Olduvai fossil skull in the official monograph in which he reassigned the specimen as *Australopithecus (Zinjanthropus).* The Olduvai find was merely a variety of Dart's fossil and was, after all, an ape, worthy only of a mention in brackets.[7]

As far as the age of the fossil is concerned, we saw in Chapter 5 that the potassium-argon method of dating has yielded dates ranging from 160 million to 3,000 million years for rocks formed in a volcanic eruption only 190 years ago, while the scatter of dates for the volcanic KBS Tuff of Lake Turkana ranged from 0.5 to 17.5 million years. The method is subject to so many separate sources of major inaccuracy that little confidence can be attached to dates stretching back millions of years. If the dating of the associated rock formation is subject to error then we are left in the dark regarding the age of its fossil contents.

One further puzzle remained regarding the area in which the Leakeys had found *Zinjanthropus* and that was the presence of stone tools. If all the fossils found so far were apes, who had made the thousands of tools which littered the Olduvai Gorge?

The answer was found in 1964, when once again the pages of *National Geographic* and *Nature* resounded with the discovery of yet another missing link. This time it was a new species of human, man the toolmaker, *Homo habilis* or handy man. Again it was the Leakeys who made the discovery.

On this occasion the find was sparse indeed, consisting only of a lower jaw with teeth, a collarbone, a finger bone, and some small fragments of skull. For the first time, a new human species was to be described on the basis of teeth and fragments alone, and in circumstances where the association of the bones as those of a single individual was conjectural—a situation very reminiscent of Dubois and Java Man.

Since 1964 *Homo habilis* has been reevaluated and it has been suggested that one of the hand bones is a piece of vertebra, that two more bones could have belonged to a tree-dwelling monkey, and that six others came from some unspecified nonhominid. But whatever the merits of the original description, the fact remains that handy man is a human—not a missing link. *Homo habilis* is calculated to have had a small brain: perhaps only half the size of the average modern human's. But, as Dr. A. J. White has pointed out, the habilines were also small in stature, so their brains were not small in relation to their body size, rather like modern pygmies.[8]

Indeed, one of the ironic aspects of the discovery of *Homo habilis* is that while Darwinists concentrate their attention on interpreting finger bones and vertebrae at Olduvai Gorge, attempting to establish the creature's credentials as a missing link, they appear to have overlooked the fact that only a few hundred miles to the east, in the forests of Zaire, are the Mbuti people who are on average only four feet six inches tall and who, in stature, brain capacity, and even way of life, are comparable to *Homo habilis.* Yet the Mbuti are modern men in every sense except that they do not watch television documentaries nor receive grants from science-funding bodies.

Other workers have continued to unearth early remains in Africa, notably Donald Johanson and his team working in the Afar region of northern Ethiopia. Johanson has discovered bones and teeth which represent up to 65 individuals, including the famous "Lucy" (*Australopithecus afarensis*) discovery which consists of 40 percent of a complete skeleton.[9]

Lucy was immediately and enthusiastically greeted as a missing link, Darwinists apparently having forgotten that it was Lucy's *Australopithecine* relatives that Solly Zuckermann found were "predominantly ape-like, and not man-like creatures" some thirty years earlier.

Lucy's apelike character was also forgotten about when she was restored to lifelike appearance for display in the Natural History Museums of London and New York and elsewhere. From her glass case Lucy peers with an intelligent gaze at visitors, her posture fully erect and humanlike, her hands and feet also short and humanlike.

This restoration must have come as something of a surprise to anatomists Jack Stern and Randall Susman of the State University of New York who, in their 1983 study published in the *American Journal of Physical Anthropology,* described the anatomy of Lucy's species *Australopithecus afarensis.* They described Lucy's hands and feet as being long and curved, typical of a tree-dwelling ape. Indeed, their paper shows that both the finger and toe bones of Lucy's species are highly curved even when compared to those of a modern ape like a chimpanzee.[10] Just why Lucy should have been restored to have humanlike hands and feet, contrary to the known anatomical facts, remains a mystery which only her restorers can explain.

Paleontologists have continued to make finds of bones and teeth in Africa, Asia, and elsewhere. But despite more than a century of energetic excavation and intense debate the glass case reserved for mankind's hypothetical ancestor remains empty. The missing link is still missing.

CHAPTER 19

Hopeful Monsters

FRENCH BIOLOGIST JEAN BAPTISTE DE LAMARCK suggested in his *Philosophie Zoologique* in 1809 that changes in environment would alter an animal's needs, that this in turn would change its behavior, and that the changed pattern of behavior would alter its physical structure. As an example, Lamarck pointed to wading birds and suggested that, "wishing to avoid immersing its body in the water, the bird acquires the habit of elongating and stretching its legs." Not only did Lamarck think that using organs made them grow (like exercising muscles), he also thought that not using them made them disappear, like the eyes of the mole.

The problem with this suggestion is that if it were true, then the weightlifter's son would be born with big muscles and the ballerina's daughter would be born dancing. In fact, say Darwinists, characteristics are inherited according to Mendel's law of inheritance: dominant genes preponderate in the offspring, not acquired characteristics.

Today Lamarck is scorned as belonging to the prescientific (that is to say, pre-Darwinian age) and the charge of "Lamarckism" is the most dreaded heresy in the evolutionists' canon. It is strange that such a distinguished biologist should be treated thus, especially since Darwin himself continually flirted with the inheritance of acquired characteristics and cited many examples, including the case of a man who was reported to have lost his fingers and later produced sons also without fingers.

As far as Darwinists today are concerned, the matter is settled and the debate is closed. Any backsliding from the straight-and-narrow line of mutation with natural selection is written off as "Lamarckism." But like so many issues in evolution theory, this one refuses to go away. For despite the often repeated claim that no one has demonstrated repeatably the inheritance of acquired characteristics experimentally in the laboratory, the fact is that numerous researchers have done just that.

In the field of botany is the work of Alan Durrant of University College of Wales, Aberystwyth, who in 1962 induced changes in the flax plant by means of different kinds of fertilizer. Durrant bred some flax plants that were larger and heavier than the parent stock and another strain that was lighter and smaller. These trends persisted when the plants were bred in successive generations. The plant breeding was carried on for more than twenty years and it was shown that when the large and small plants were crossbred, the offspring exhibited the Mendelian pattern of inheritance, proving that the change is genetic.[1]

The results have been replicated by J. Hill at the Welsh Plant Breeding station, where a permanent change has been effected in the tobacco plant, *Nicotiana rustica*. In the case of the tobacco experiments, the flowering time was also changed.[2] Christopher Cullis has reviewed all the work and has suggested a model to explain the induced changes in terms of molecular genetics.[3]

It may be said that these experiments are relevant to plants but not to animals. However, there is also experimental evidence from the animal world. As long ago as 1918, Guyer and Smith ground up the eye lenses of rabbits and injected the resulting substance into birds. When serum made from the birds' blood was injected into rabbits, their offspring were born with small or defective eyes, or with none at all, and the defects continued through the succeeding nine generations. The reason for choosing eye-lens tissue was that it is known to provoke immune reactions.

The specific reason that Darwinian geneticists reject any form of Lamarckism is their belief that the genes are unalterably separate from the cells of the body and that there is no route by which changes could be communicated to them from outside. This belief was first enunciated by August Weisman in his 1893 book *The Germ Plasm: A Theory Of Heredity*. It was restated as recently as 1970 by no less an authority than Francis Crick who, with James Watson, deduced the structure of the DNA molecule and the code by which it transmits genetic information. Crick said that genetic information could travel from DNA to protein but not from protein to DNA.[4]

Both Weisman and Crick have been shown to be mistaken by the work of Howard Temin at Wisconsin University in 1971. Temin discovered that viruses can transport genetic material into host cells and embed it in the host DNA where it will later replicate itself using the host cell's factory facilities for synthesizing proteins. In order to perform this biological confidence trick, the viruses manufacture a special enzyme which Temin called *reverse transcriptase*. For this discovery he received a Nobel prize in 1976.[5]

Having found a two-way channel of communication between the genes and the outside world, science still lacked a mechanism by which the demands of the environment could directly affect the germ cells: how, say, the wading bird who is constantly stretching could transmit to his genes his desire for longer legs to stay dry. Three years after Temin stepped onto the podium in Stockholm to shake hands with King Gustav, a young Australian biologist named Ted Steele proposed just such a mechanism. In 1979 Steele proposed that mutations could occur in body cells, be copied to other body cells by viruses, and finally be transmitted by viruses to the germ cells of the sperm in men or egg in women, and so become inheritable.[6]

The next problem was to design an experiment that would test Steele's idea. A colleague of Steele's, the Canadian Reg Gorczynski, neatly solved the problem by constructing an experiment merely by adding a new twist to the famous experiment of Peter Medawar.

Medawar won a Nobel prize for showing that immune tolerance can be acquired from outside. His original experiment was concerned with a phenomenon that has become familiar to everyone in an age of organ transplants—that of rejection of tissue. The body's immune system will reject any cells that are not genetically identical and hence which it identifies as alien. Medawar showed that if alien cells are injected into a newborn mouse then, later in life, it will accept a skin graft from the same source. His experimental success was dramatically depicted in newspapers and magazines around the world by pictures of black mice having patches of white skin successfully grafted on and white mice equally at home with black patches.

Reg Gorczynski set out to duplicate Medawar's experiment, but with the goal of finding out if the established tolerance was inheritable. His experiment showed that 50 percent of the offspring of the tolerant mice were also tolerant in the next generation and the grandchildren were tolerant in between 20 percent and 40 percent of individuals.[7]

On the face of it, this experiment successfully demonstrated the genetic inheritance of *acquired* immunity. It is fair to add that a team of distinguished scientists including Medawar himself and Professor Leslie Brent of St. Mary's Hospital Medical School attempted to repeat Steele and Gorczynski's results and were unable to do so. The position at the moment is that the jury is still out. But regardless of the outcome of this particular experiment, it is no longer possible for Darwinists to assert that outside agencies cannot communicate genetic changes via the mechanism of DNA replication. They can.

Additional, and very suggestive evidence, has come from two series of experiments conducted in recent years in the United States. The first was conducted by British biologist Dr. John Cairns and two colleagues at Harvard University in 1988. The second, a repeat of the Cairns experiments with tighter controls and extended objectives, was carried out by Dr. Barry Hall of Rochester University in 1990. The experiments were conducted on bacteria, principally the species *Escherichia coli*. What they demonstrate is that when the bacteria is deprived of certain essential nutrients such as the amino acids tryptophan and cysteine, they are capable in this extremely hostile environment of giving rise to descendants able to synthesize their own nutrients. What is taking place, believe Cairns and Hall, is that the bacteria are mutating and that the mutation is not random but internally directed by the needs of the organism in the direction of being able to synthesize the necessary nutrients.[8]

If these latest experiments are confirmed, it will almost certainly mean that we must look again much more closely at some form of Lamarckism. However unlikely it seems and however difficult it proves to obtain experimental confirmation, it looks increasingly probable that, in some unknown way, individuals cannot only adapt to their environment or way of life but can also sometimes pass on that adaptation to their offspring.

The aim of this chapter is to summarize the main alternatives to neo-Darwinism, of which some form of Lamarckism is possibly the principal contender. But it is by no means the only serious alternative proposal. Others include evolution by sudden jumps, the cause of which is uncertain, proposed by Richard Goldschmidt, professor of genetics at the University of California at Berkeley; the origin of life from space, as proposed by astronomer Fred Hoyle; and a specialist variety of the extraterrestrial hypothesis advanced by Francis Crick.

There are even some paradigm-shattering and entirely novel approaches to biology such as the theory of formative causation proposed by Rupert Sheldrake.

Some of neo-Darwinism's most important supporters have defected from the cause in recent decades and have espoused various heretical alternative ideas. The most prominent biologists to defect from the synthetic theory since the Second World War have been Richard Goldschmidt, who described the function of the chromosome, and C. H. Waddington, maverick professor of biology at Edinburgh University. The most recent heretics have included Harvard's professor of paleontology Stephen Jay Gould, his fellow paleontologist Niles Eldridge, and British astronomer Fred Hoyle. All have dared to challenge the received wisdom of uniformitarian rates of change, and slow gradual microscopic mutation coupled with blind chance.

Goldschmidt's concern has been to account rationally for the puzzling gaps in the fossil record by accepting them as real, rather than as inconvenient obstacles to an otherwise elegant theory. Goldschmidt coined the poignant and graphically descriptive phrase "hopeful monster" to describe his heresy. His idea is simply that perhaps evolution proceeds by large jumps (known in the trade jargon of evolution as "saltations"). Perhaps macroscopic mutations have occurred such that one day a reptile laid an egg from which hatched the first bird.[9]

This is actually no less probable than the idea that bears might mutate into whales. But the trouble with hopeful monsters is that they create a problem of exactly the same magnitude as the one Goldschmidt is trying to solve. To get from an ancestral reptile to a winged birdlike creature by conventional neo-Darwinist micro-mutations would take 100 or 1,000 or perhaps even 10,000 individual steps, each step representing a generation. Taking Julian Huxley's rate of mutation (once in every million births) as a rough guide, then one would expect many millions—probably billions—of transitional individuals to have lived, at least some of which would be represented in the fossil record.

However, if viable macromutations occur (and there is no evidence for them, just as there is no evidence for beneficial micro-mutations) then most of them would be disadvantageous (perhaps fatal) to their carriers. Wings might conceivably be of assistance to a small lightweight reptile, but would be of little help to a 20-foot sea-going crocodile or a 200-foot brontosaurus. So there would have to be just as high a ratio of unsuccessful hopeful monsters to successful hopeful monsters as there would be transitional micromutations to stable species.

More simply, the fossil record would be littered with the bodies of one-shot macromutations that did not work. For every macromutation like the hypothetical bird, there would be millions of one-legged crocodiles or aardvarks with wings. In fact, no one has recorded finding a single such failed monster. For Goldschmidt's idea to be correct, nature would have to have had a 100 percent success record. The difficulty this creates is that if nature possesses a mechanism that makes trial and error unnecessary, then the entire apparatus of random mutation and natural selection goes out the window.

Stephen Jay Gould and Niles Eldridge of Harvard have proposed a theory of "punctuated equilibrium," in order to account for the lack of fossil remains of transitional species. They have suggested that evolution is not a constantly occurring phenomenon. Species may have remained stable for long periods of geological history, leaving many fossil remains, and the periods of evolutionary change, when they came, did not last for long. This would account for the lack of transitional fossils.[10] The difficulty with punctuated equilibrium is that it is wholly speculative and has been introduced simply to account for the lack of fossils that ought to exist in the neo-Darwinist theory.

One professional biologist has gone so far as to question the very foundation of the life sciences, and the belief in a mechanistic, reductionist basis for living things. Rupert Sheldrake has audaciously confronted the kind of biological problems that most scientists prefer to ignore and has predictably been derided by some of his colleagues.

Sheldrake has sought answers to the unexplained mysteries thrown up by research that are usually ignored by science. When a laboratory rat has learned a new trick in one place, other rats elsewhere seem to be able to learn it more easily. When new chemical compounds, such as antibiotics, are made for the first time they are difficult to crystallize but the more often they are made the easier their crystals form. When some birds first learned to open milk bottles on the doorstep, birds all over the country suddenly learned the same trick.

Sheldrake's solution to these puzzles, proposed in his 1983 book *A New Science of Life*, is that organisms and species can learn, develop, and adapt through a process he calls morphic resonance.[11] Living things are built on universal templates called morphogenetic fields, says Sheldrake, through which some plants and animals are able to regenerate damaged or missing tissue, as the salamander or starfish can grow a new leg.

Intriguingly, Sheldrake insists that morphic resonance is accessible to empirical study and has proposed numerous experiments to test his theory, some of which were carried out through public television broadcasts in Europe and America in 1984. One such experiment was a picture containing a "hidden" image that requires a certain amount of concentration to "get." Would showing the image to millions of people on TV make it easier for people who could not have seen the broadcast to "get" the hidden image?

A significant positive effect was obtained in Finland, France, Germany, Italy, Sweden, Switzerland, and Yugoslavia, with an overall increase of 32 percent in people "getting" the picture. Strangely, in America and Canada there was no overall significant effect at all, so the jury is still out on an experimental confirmation of morphic resonance. Meanwhile, further experiments are planned.

The reaction of orthodox science to Sheldrake's book can per-haps serve as the best modern illustration of the fate that awaits anyone who challenges the orthodoxy of neo-Darwinism. The edi-tor of *Nature* magazine, John Maddox, ran an editorial calling for the book to be burned—a sure sign, many will think, that Sheldrake is onto something important.

Some areas of research imply that biologists may have got some of their most cherished principles spectacularly wrong. For example, although universally accepted today, the principle that living things arise only from living things and never spontaneously is by no means as securely proved as most Darwinists imagine.

It was this question—at the heart of biological controversy for centuries—that Louis Pasteur settled with his famous experiment in which he prepared several sterile "swan-neck" flasks of nutrient medium and left some of them open but kept some sealed. Molds appeared in the open flasks but not in the sealed flasks, proving once and for all that the idea of life spontaneously coming into being was false, and also lending strong support to the idea that the molds were caused by microorganisms in the air.

Recently, American biologist Gerald Geison was able to gain access to Pasteur's notebooks and has translated and published them.[12] Geison looked up Pasteur's private observations about the famous experiment and discovered that, in fact, Pasteur *did* find evidence that life flourished in his sealed sterile jars, but he chose to ignore it. He wrote, "I did not publish these experiments, for the consequences it was necessary to draw from them were too grave for me not to suspect some hidden cause of error in spite of the care I had taken."

In other words Pasteur's attachment to his theory was too strong to be overcome by empirical evidence even from his own experi-ments. More significantly, he chose to keep the contradictory evi-dence secret.

Other researchers—most notably Wilhelm Reich—have in-sisted that there *is* experimental evidence which shows that elemen-tary life forms such as protozoa assemble themselves spontaneously from decaying organic material.[13] Reich's 1950s experiments were replicated in 1987 by Dr. Robert Dew who published remarkably detailed color photographs of the process apparently taking place.[14] Interestingly, Reich, too, asserted that living things are informed by and develop through the agency of a bioenergy field.

The ideas of Fred Hoyle and his fellow astronomer Chandra Wickramasinghe on an extraterrestrial origin of life are guaranteed to liven up an otherwise dull winter's evening at your local bar. Surprisingly, their proposal is not new. In 1908 the distinguished Swedish chemist Svante Arrhenius suggested, in his book *Worlds in the Making*, that living spores could be driven through space by the pressure of light from the stars.

Hoyle and Wickramasinghe's proposal is based on their belief that interstellar space is filled with clouds of dust consisting mainly of cellulose or similar sugarlike organic material. The comet Kohoutek was examined spectroscopically on its near approach to Earth in 1973 and was found to contain at least two organic molecules—methyl cyanide and hydrogen cyanide—along with rock dust, polysaccharides, and related organic polymers, all of which are possible building blocks of life. The two astronomers' idea also involves the idea of the Earth colliding with a comet at some time in the past.[15]

The authors say, "The Earth could have acquired all of its volatiles—including all the oceans—from such collisions [with comets]. And, of course, the presence of organic prebiotic chemicals such as we have discussed would have led to a vast input of life-forming materials to the Earth."

Hoyle has made a powerful case for these ideas. In recent years several microorganisms have been discovered that can withstand the extremely hostile conditions of space. The bacterium *Micrococcus radiophilus* can survive exposure to X rays at doses that would kill humans, while *Pseudomonas* has been found living quite happily in the core of an American nuclear reactor. Bacteria of the species *Streptococcus mitis* were inadvertently sent to the Moon in the unmanned Surveyor III in 1967 and were "rescued" still alive two years later by the crew of Apollo 12 who brought back Surveyor's TV camera. The organism had been subjected to very low pressure and temperatures of minus 100 degrees Celsius.

Even more significantly, in 1981, Hans Dieter Pflug tentatively identified microorganisms closely resembling the bacterium

Pedomicrobium and a virus resembling influenza, inside a meteorite that fell in Australia in 1969. (As well as being among the first to identify extraterrestrial organisms, it was Pflug who identified the oldest fossil on Earth, the 3,800-million-year-old *Isosphaera* organism.)[16]

Francis Crick has made a further proposal. In his book *Life Itself* he, too, suggests an extraterrestrial origin for life but believes that it is unlikely that organic molecules of any complexity could survive drifting in interstellar space. He suggests instead that life in microscopic form may have been sent to other planets by alien beings in suitable protective vessels; that life is like a message in a bottle.[17]

Pflug's identification of microorganisms in a meteorite was treated with deep skepticism by most of his fellow geologists. But in 1996 the idea of life from space received a boost from none other than the National Aeronautics and Space Administration (NASA). Biologists working for NASA in Houston and at Stanford University claimed to have found traces of the activity of complex organic chemistry associated with microorganisms fossilized in a meteorite discovered in 1984.

Known as ALH84001, the 4.2-pound rock is believed to have originated on Mars and was found in the Allan Hills ice field in Antarctica. In 1993 the meteorite was identified as Martian in origin by its characteristic mineral signature. And researchers discovered globules of carbonate minerals which seem to be the remains of microorganisms closely resembling terrestrial fossils of bacteria. The researchers also found quantities of polycyclic aromatic hydrocarbons, which are a by-product of organic decay, as well as iron sulphide and magnetite minerals which are generated by living bacteria on Earth.

If microscopic life originated elsewhere in the universe, then it is still necessary to account for the evolution of life from the microscopic to the macroscopic level. Extraterrestrial origin would assist Darwinists to the extent that it relieves them of accounting for the spontaneous synthesis of self-replicating molecules on Earth; although of course, it is still necessary to account for their emergence on the planet of origin. This idea also assists Darwinists to the extent that it enables one to conjecture a planet with conditions different from those on Earth, and more favorable to the formation of life, perhaps even with uniformitarian conditions of planetary development and with the requisite billions of years of time available, which we now know may not have been available on Earth.

At root, however, the same basic questions arise wherever life is said to originate: What nonliving mechanism can have given rise to the first self-replicating cells and how? And what was the mechanism of evolution from the cellular level to the present-day plant and animal kingdoms?

Was it random mutation with natural selection? Punctuated equilibrium? The hopeful monster? Morphogenetic fields? Or something else entirely?

CHAPTER 20

The Facts of Life

Zoologist Bernard Heuvelmans once observed that just because a country is on the map, it doesn't mean that we know all about its inhabitants. Science has achieved miracles in the elucidation of the most complex microscopic structures and the furthest galaxies, yet there are deeper unsolved mysteries in the average suburban garden.

The swallowtail butterfly begins its life cycle by emerging from an egg as a caterpillar, enters a pupal or chrysalis stage, and re-emerges as the familiar winged insect. While inside its pupa, however, the caterpillar undergoes a metamorphosis whose nature is not understood at all. The body of the caterpillar dissociates completely into an amorphous cellular liquid referred to as a "soup." The soup then reorganizes itself into the structure of a butterfly.

To say that this process is not understood is not merely to say that television cameras have not so far been lucky enough to catch it on film. It means that no stage or aspect of this physical process can be accounted for or even guessed at with our current knowledge of chemistry, physics, genetics, or molecular biology, extensive though they are. It is completely beyond us. We know practically nothing about the plan or program governing the metamorphosis, or the organizing agency that executes this plan.

In attempting to gather the strands of evidence from the natural world that might point the way to an alternative view of evolution, there seem to be three key kinds of observation, three persistently recurring themes that are crying for answers: the unerring accuracy of nature, her lack of trial and error; the presence of a systematic program above the cellular level, controlling somatic development; and the overwhelming probability that environmental factors can in some unknown way directly affect the genetic structure of the individual.

The nonexistence of transitional types (including failed monsters) in the fossil record and in the contemporary animal kingdom shows that nature goes unerringly to its target. The human eyelid exactly covers the human eye. The process that made the eyelid grow stopped when the eyelid was the right size. It cannot be maladaptive to have an eyelid a little longer than needed—yet no creature has such an "imperfection" in this anatomical detail or any of the myriad other details.

This is merely among the obvious examples of a universal phenomenon that we take for granted. A child's second teeth are adult sized even though they appear at age seven when the child's jaw is not yet full grown; little orange trees have little oranges and large orange trees have large ones; the individual parts of any organism, from a tadpole to an elephant, are all in the correct relative scale.

At the turn of the century Henry Williams, of the University of North Carolina, made an illuminating discovery when he pressed sponges through a cloth until they were dissociated into individual cells. The cells spontaneously came together again and formed new sponges on their own initiative. In 1963 T. Humphreys confirmed and enlarged on Williams's discovery. Johannes Holtfreter of the University of Rochester took up Williams's experiments after the Second World War and found that the cells from the embryos of vertebrates will also reassemble themselves when dissociated. In 1952 A. A. Moscona, of the University of Chicago, tried similar experiments with the tissues of chicks and mice. He found that dissociated kidney cells not only reassembled into kidney tubercles but also began to secrete kidney enzymes. Similarly, liver cells will reassemble into structures resembling the intact organ and carry out the liver function of accumulating glycogen. Heart cells, almost incredibly, coalesced into rhythmically contracting tissue.[1, 2]

This sort of behavior is also inexplicable at present, not, I suspect, because of some matter of detail, but because there is something big happening of which we know nothing as yet. It is not a matter of cells merely being attracted together like carbon atoms. These cells have a joint function which they cannot possess individually—like the heartbeat of heart cells. There is a program being executed. How is it coded? Where are the instructions?

The many resoundingly pointless breeding experiments with the fruit fly *Drosophila* did yield one highly illuminating discovery. To use the experimenters' terminology, the fly possesses a mutant recessive gene (that is, one which normally plays no part in reproduction) which, if present in both parents, results in an offspring that is eyeless. If a stock of such eyeless flies is bred, then their offspring can only be eyeless too. Yet within a few generations offspring appear which do have normal eyes.

It would be absurd to imagine that nature has repeated in a few months what is supposed to have taken millions of years to occur: the origination of an eye by chance mutation. The orthodox explanation of this phenomenon is that the other genes have somehow "deputized" for the missing gene by a recombination.

The significant point about the eyeless fly is that it again demonstrates some kind of global program control in action. The fly's genetic mechanism "knows" that it lacks an important gene and is able to take effective "action" to compensate. The question is where does this program reside and how is it invoked and executed?

As long ago as 1895, German biologist Hans Driesch performed an experiment that caused him to develop a whole new philosophy of biology. While working at the Zoological Research Station at Naples, Driesch experimented on a sea urchin egg. He killed half the egg but discovered that the remaining half developed into a perfectly normal embryo, except that it was half the normal size. A rather similar experiment was conducted by B. I. Balinsky in 1951 when he transplanted tissue from the embryonic optic nerve of a large species of amphibian to the embryo of a smaller but related amphibian. The result was a perfectly formed eye with all parts in proportion, but intermediate in size between the two animals.[3]

Once more, in both these cases, there is some kind of global supervisory function being exercised which seems to be "aware" of an overall plan.

The example of the parallel evolution of the placental mammals of Europe and the marsupial mammals of Australia, described in Chapter 17, is further evidence for some overriding principle at work. Darwinists believe that a single shrewlike ancestor has independently evolved into carbon copies of wolves, cats, rats, and a dozen other mammals on widely separated continents, simply by virtue of their common lifestyles.

But it is obvious that if the shrewlike creature of the Cretaceous really is the ancestor of both marsupials and placentals, then its evolutionary "trajectory" has been strictly circumscribed by natural laws, just as the flight of a cannonball is circumscribed by gravity. The repertoire of options open to evolution has been dictated by a strategic plan or program. Where does that program reside? How is it executed? What is the "gravity" of evolution? Natural selection is an inadequate explanation.

Earlier on I referred to computers and their programs as a fruitful source of comparison with genetic processes since both are concerned with the storage and reliable transmission of large quantities of information. Arguing from analogy is a dangerous practice, but there is one phenomenon connected with computer systems that could be of some importance in understanding biological information processing strategies.

The phenomenon has to do with the computer's ability to refer to a master list or template and to highlight any exceptions to this master list that it encounters during processing. This "exception reporting" is profoundly important in information processing. For instance, this book was prepared using a word-processing program that has a spelling checker. When invoked, the spell-checker reads the typescript of the book and compares each word with its built-in dictionary, highlighting as potential mistakes those it does not recognize. Of course, it will encounter words that are spelled correctly but are not found in a normal dictionary—such as "deoxyribonucleic acid." But the program is clever enough to allow me to add the novel word to the dictionary, so that the next time it is encountered it will be accepted as correct instead of reported as an exception—as long as I spell it correctly.

In other words, the spell-checker isn't really a spelling checker. It has no conception of correct spelling. It is merely a mechanism for reporting exceptions. Using these methods, programmers can get computers to behave in an apparently intelligent or purposeful way when they are really only obeying simple mechanical rules. Not unnaturally, this gives Darwinists much encouragement to believe that life processes may at root be just as simple and mechanical.

In cell biology there are natural chemical properties of complex molecules that lend themselves to automatic checking and excepting of this kind. For example many molecules are stereospecific—they will attach only to certain other specific molecules and only in special positions. There are also much more complex forms of exception reporting, for instance as part of the brain's (of if you prefer, the mind's) cognitive processes: as when we see and recognize a single face in the crowd or hear our name mentioned at a noisy cocktail party.

In the case of the spell-checker, the behavior of the system can be made to look more and more intelligent through a process of learning if, every time it highlights a new word, I add that word to its internal dictionary. If I continue for a long enough time, then eventually, in principle, the system will have recorded every word in the English language and will highlight only words that are indeed misspelled. It will have achieved the near-miraculous levels of efficiency and repeatability that we are used to seeing in molecular biological processes. But something strange has also been happening at the same time—or, rather, two strange things.

The first is that as its vocabulary grows, the spell-checker becomes *less* efficient at drawing to my attention possible mistakes. This unexpected result comes about in the following way. Remember, the computer knows nothing of spelling, it merely reports exceptions to me. To begin with, it has only, say, 50,000 standard words in its dictionary. This size of dictionary really only covers the common everyday words plus a modest number of proper nouns (for capital cities, common surnames, and the like) and doesn't leave much room for unusual words. It would, for instance, include a word like "great" but not the less-frequently used word "grate."

The result is that if I accidentally type "grate" when I really mean "great," the spell-checker will draw it to my attention. If however, I enlarge the dictionary and add the word "grate," the spell-checker will ignore it in future, even though the chances are that it will occur only as a typing mistake—except in the rare case where I am writing about coal fires or cookery.

One can generalize this case by saying that when the dictionary has an optimum size of vocabulary, I get the best of both worlds: it points out misspellings of the most common words and reports anything unusual which in most cases probably will be an error. (Obviously to work at optimum efficiency the size of dictionary should be matched to the vocabulary of the writer). As the dictionary grows in volume it becomes more efficient in one way, highlighting only real spelling errors, but less efficient in another: it becomes more probable that my typing errors will spell a real word—one that will not be reported—but not the word I mean to use. Paradoxically, although the spell-checker is more efficient, the resulting book is full of contextual errors: "pubic" instead of "public," "grate" instead of "great," and so on.

It requires a human intelligence—a real spelling checker, not a mechanical exception reporter—to make sure that the intended result is produced.

I said two strange things have been happening while I have been adding words to the spell-checker. The second is the odd occasion when the system has highlighted a real spelling mistake to me—say, "problem" instead of "problem"—and I have mistakenly told the computer to add the word to its dictionary. This, of course, has the very unfortunate result that in future it will cease to highlight a real spelling mistake and will pass it as correct. The error is no longer an exception it is now a dictionary word.

Under what circumstances am I most likely to issue such a wrong instruction? It is most likely to happen with words that I type most frequently and that I habitually mistype. Anyone who uses a keyboard every day knows that there are many such "favorite" misspelled words that get typed over and over. Once again, only a real spelling checker, a human brain, can spot the error and correct it.

The reason that the computer's spell-checker breaks down under these circumstances is that the simple mechanisms put in place do not work from first principles. They do not work in what electronics engineers call "real time" (they are not in touch with the real world) and do not employ any real intelligent understanding of the tasks they are being called on to perform. So although the computer continues to work perfectly as it was designed to, it becomes more and more corrupted from the standpoint of its original function.

I believe that this analogy may well have some relevance to Darwinists' belief that biological processes can at root be as simple as the spell-checker. It is easy to think of any number of simple cell replication mechanisms that rely on exception reporting of this kind. I believe that if biological processes were so simple, they too would become functionally corrupt unless there were some underlying or overall design process to which the simple mechanisms answer globally, and which were capable of taking action to correct mistakes. This is the mechanism that we see in action in the case of the "eyeless fly," *Drosophila*; in Driesch's experiment with the sea urchin and Balinsky's with the eyes of amphibians; and in the "field" that governs the metamorphosis of the butterfly or the reconstitution of the cells of sponges and vertebrates.

Darwinists believe that the only overall control process is natural selection, but the natural selection mechanism could not account for the cases referred to above. Natural selection works on populations, not individuals. It is capable only of tending to make creatures with massively fatal genetic defects die in infancy, or to make populations that are geographically dispersed eventually produce sterile hybrid offspring. It is such a poor feedback mechanism in the sense of exercising an overall regulating effect that it has failed even to eliminate major congenital diseases. Natural selection offers only death or glory: there is no genetic engineering nor holistic supervision of the organism's integrity. Yet we are asked to believe that a mechanism of such crudity can creatively supervise a program of gene mutation that will restore sight to the eyeless fly.

This is plainly wishful thinking. The key questions remain: What is the location of the supervisory agency that oversees somatic development? How does it work? What is it's connection with the cell structure of the body?

Whether they are Darwinists or vitalists, biologists have begun to talk in terms of "morphogenetic fields." D. J. Pritchard, a Darwinist geneticist from Newcastle University, wrote in 1990:

There is a great deal of evidence that organs and organisms have an awareness of their "wholeness" (Dalq, 1951; Lillie, 1927; Spemann, 1924) such that when a portion of the whole is lost, steps are taken to replace it. For example salamanders will regenerate their limbs (French, Bryant and Bryant 1976; Wallace, 1981); if a sponge is disaggregated into single cells these will reaggregate to form a perfect sponge (Humphreys, 1963). Embryologists recognise "morphogenetic fields" which have spatial unity with respect to the organization of their constituent parts. If a field is divided into two a complete structure can form in each half independently of the other. Our own retinas began as the two halves of an initially single retinal field. If division of the retinal field fails the result is a single, central eye, a condition known as cyclopea. What evolution has created within the bodies of animals are integrated, self-organizing systems which are not just defined by their component parts, but actually define those components.[4]

These tantalizing glimpses of the unity of organic structures are as far as our present experimental knowledge takes us. Only further experiment and a certain amount of luck can provide the hard data that will solve fully these baffling questions and we must await the acquisition of new facts. In the absence of concrete answers, I would like to offer some speculations.

To begin with, we have a working hypothesis in Ted Steele's proposal that viruses are able to replicate mutations in somatic cells and transfer them to sexual cells, where they become inheritable. The next question to be asked is, What kind of cellular changes might be induced in somatic cells? And, exactly how might they be induced?

C. H. Waddington—an unusual combination of an academic with an anarchic sense of humor—has essayed just such a mechanism. It must be said that Waddington dreamt up this mechanism in a light-hearted vein simply to infuriate orthodox neo-Darwinists (especially Jacques Monod of the Pasteur Institute who had accused Waddington of being a Lysenkoist—an even worse crime than Lamarckism). In his essay "How Much Is Evolution

Affected by Chance and Necessity?" Waddington includes a massive footnote outlining his idea.[5] In a crude and simplified form
it is this. It has been established that important parts of the DNA
molecule are repeated many times in the chromosomes—rather
like back-up tapes. Just like back-up tapes, these replicate versions may vary slightly. There is also another set of tapes in the
form of mitochondrial genes, which are further structures in the
cell. All these genes are closely involved with the important metabolic processes that go on within the cell. So it is not inconceivable that the rates of multiplication of slightly differing genes
would be influenced by the particular metabolic circumstances
reigning in the cell in question. And it is not inconceivable that
the imposition of certain metabolic conditions on an organism
might change the proportion of variant forms of gene within the
population (of all the back-up copies) to be passed on to the next
generation. The effect of this would quite simply be the direct
inheritance of an acquired character.

Put more simply, the metabolic stresses placed by an individual
on his cellular structure might determine which tape is selected
from the library for duplication.

Imagine, for example, a very athletic woman stimulating the
metabolism of her cells in such a way that replicate DNA sequences
coding for physical agility are promoted preferentially and as a result she gives birth to athletic daughters.

Waddington called his idea an "outrageous speculation." What
he may not have known when he dreamt it up in 1974 is that he
had only to account for differential multiplication of the DNA replicates in ordinary body cells: Steele's viruses could replicate the
chosen DNAs to the sexual cells through "reverse transcription."
This could make his suggestion hundreds if not thousands of times
less outrageous and more probable.

The hypothetical example given above would be an example of
a physical behavior affecting somatic cells. Are other forms of influence possible? The answer appears to be that psychological states
may also affect somatic cells. Epidemiologists believe they have
identified a "cancer personality"; a set of individual, character traits
which, if possessed mainly or wholly by one individual may predis-
pose that person to cancerous illness: that is, the faulty replication

of somatic cells. If it is true that personality factors can affect cell biology, and if viruses can copy genetic mutation from somatic cells to sexual cells, then it follows that personality factors could in principle be inheritable. To carry this speculation a step further, some of the personality traits that compose the "cancer personality" are psychological rather than physical (for example excessive anxiety). This raises the possibility that purely psychological factors could be translated into both somatic and ultimately genetic factors, that the content of an individual's consciousness could affect his or her body and the bodies of any offspring.

Presumably, the metabolism of the "cancer personality" is actually different from that of a noncancer personality in some distinctive way; for instance, anxiety may alter the balance of some hormone or enzyme which ultimately results in alteration to somatic cells. If so, the nature of those differences may hold an important key for biology.

One further possibility—perhaps a rather disturbing one—remains to be explored. For more than fifty years it has been recognized that, at the nuclear level, our solid world dissolves into a cloud of fuzzy probabilities. Until recently, lip service was paid to the principle of uncertainty in physics, but no serious scientist would care to admit that he had designed an experiment taking himself into account. Now a concrete experimental result has been obtained which clearly shows the influence of the observer at the quantum level.

Wayne Itano and colleagues at the National Institute for Standards and Technology in Colorado reported an atomic experiment in *Physical Review* in March 1990 in which the result was determined by the observer.[6]

The NIST experiment involved heating with radio waves a container of beryllium atoms and measuring the ratio of isotopes formed. The radio pulses convert the atoms from one isotope to another. The researchers used a laser beam to display the results since it would cause atoms in their original state to emit light but not atoms in the altered state.

What they found was that the more measurements they made with their laser beam, the greater the number of atoms that remained unaltered. The very act of observing the atoms stopped them from changing state, regardless of the effect of the radio pulses. This is not simply a matter of the laser beam preventing the experiment from progressing or directly interfering with the changes in atomic state. The explanation is that observing a particle causes it to collapse from a fuzzy probabilistic cloud into a definite mass at a definite point in space and time, as predicted by quantum mechanics.

The question this experiment raises is, if merely observing an event causes changes to occur at the atomic level, and if genetic coding is controlled by atomic structures, can genetic mutation be caused by direct influence at the quantum mechanical level?

Is it even conceivable that, as Hans Driesch conjectured, "the mind may carry out a morphogenetic action at a distance"? Can we wish for wings and get them? Probably not. Does a healthy mind promote a healthy body? Almost certainly. Is there anything in between? Who knows?

Thomas Huxley, Darwin's champion, observed that the great tragedy of science is the slaying of a beautiful idea by an ugly fact. Darwin's original conception was a beautiful idea. It seemed to offer an elegantly economical solution to the greatest mystery of all: the origin of life on Earth and the descent of humankind. Sadly, it has received too many mortal blows from the ugly facts of scientific enquiry to remain viable.

The prospect of facing the future without neo-Darwinism is not an attractive one. Its demise will leave a yawning gap in the life sciences and historical geology with no obvious successor theory; a hole that it is impossible to imagine being filled by any current competitor. How has life evolved if not by chance?

It is the customary fate of one who delivers the fatal stroke to be called upon to replace the deceased theory with a better one. This is thoroughly illogical, quite unfair, and perfectly understandable. While I do not possess an alternative theory in my back pocket, I may at least suggest what kind of new theory it might be, and where it might be found.

The alternative mechanisms so far discussed in this chapter have in common that they all approach the problem from the accepted premises of classical science—indeed in a way that Darwin himself might approach it if he were alive today and possessed of today's body of scientific knowledge. I have a deep-rooted suspicion, however, that the real solution may be found in adopting quite a different approach; the natural phenomena that may well provide an explanation of the origin of species are at present so imperfectly understood that they have baffled those physicists that have bothered to examine them at all, and have been almost entirely ignored by biologists.

For most of this century, physics has had to accept the indignity of a principle of uncertainty. Physicists have been compelled to drop their neat logical picture of the universe as a great machine, and their unambiguous, clockwork model of the atom. In place of these certainties, physical scientists have been obliged to put intangible, unimaginable abstractions. Instead of billiard-ball particles like electrons, there are probability waves. Instead of matter composed of particles and energy composed of waves, there is light made of particles, and objects made of matter waves. In this surreal subatomic world, matter has ceased to have any solid form and has no more than a tendency to exist.

While these turbulent events have been taking place in the physics faculty, down the corridor in the biology department it has been business as usual. Biologists have made remarkable discoveries, but they all have the familiar nineteenth century hallmarks of clockwork certainty. Deducing the structure of the DNA molecule is a brilliant scientific achievement, but the blue and red ping-pong balls of the molecular model remain frustratingly incapable of telling us what life is.

Using the mechanistic, reductionist approach of Victorian science, biology has not so much explained life as explained it away. The body is a machine, a matter of chemistry and electricity. Thought is merely a by-product of the computerlike brain which pulls the body's levers. Evolution is no more than a marriage of chance and chemistry. There is no ghost in the machine: human *is* the machine. It is out of this Frankenstein approach that neo-Darwinism was born and is sustained: by the science of Mendel and Kelvin, rather than that of Heisenberg and Planck.

As we near the end of the twentieth century, I believe that biology, too, will be compelled to drop its mechanistic approach and recognize that chemistry and statistics alone will not explain the nature of life. The absurd and baffling world of the nuclear particle

is beckoning those in the life sciences as it beckoned physicists decades ago. Biologists are, as it were, hesitating on the shores of an
unexplored continent. What they will find when they venture inland is impossible to say. But it is possible to gain some clues from
the discoveries that have been made by their colleagues from the
physics laboratory who set off some fifty years ago and have a substantial head start.

If the new physics has a central idea to sustain it, it is that of
wholeness. In 1935 Albert Einstein, Boris Podolsky, and Nathan
Rosen presented their colleagues in physics with a baffling conun-
drum. Trying to answer the question of whether quantum mechanics
really tells us anything about the nature of the physical world, the
three physicists proposed a thought experiment with astounding
consequences. They showed theoretically that atomic events which
appear to us as separate must in fact be connected in some unknown way. And, moreover, that such events can communicate information to each other instantly—faster even than the speed of
light which is thought to be a limiting velocity in the physical world.
The three physicists predicted that whatever happened to a nuclear
particle would be reflected in the behavior of its twin particle in a
closed system, regardless of where they were. Even if they were
billions of miles apart, a change in the momentum of one particle
would be instantly mirrored in its twin—as though the particles
were able to communicate their experience instantaneously.

Einstein, who doubted that quantum physics gave a real description of real events, thought it more likely that the twin particles were behaving in a way that merely appeared to be coordi-
nated in a cause-and-effect manner because they were both obeying some third, hidden factor affecting them both, a factor known
to physics as a local hidden variable. He, and most physicists, pre-
ferred this explanation because they do not like to have to draw
upon any form of inexplicable action-at-a-distance. In any case, it
was thought that even if the extraordinary connectedness of nuclear
particles was real, it was an effect which existed only at the nuclear
level—not in the real world of tables and chairs and certainly not
in the realm of biology.

In recent decades a number of research groups have conducted
physical experiments which have confirmed the unlikely predic-

tion of the Einstein-Podolsky-Rosen paradox. In 1972 Stuart Freedman and John Clauser at Berkeley performed an experiment which confirmed that photons—the quanta of light—really are mysteriously correlated. It is no mere philosophical contrivance to get physicists out of a conceptual difficulty; this wholeness or hidden connectedness is real. Even more significant, its effects can be felt at the macroscopic level, at the level of the everyday world including that of biology.

David Bohm, professor of physics at Birkbeck College at the University of London, has written of this connectedness in his book *Wholeness and the Implicate Order.*[7] Bohm sees the cosmos as a connected whole which he terms the implicate, or enfolded, universe. The fragments of it that we perceive with our human minds and senses he terms the unfolded or explicate world. We see and understand only a tiny fraction of the underlying connected whole—the tip of the cosmic iceberg, as it were.

So far, few biologists have abandoned the conventional viewpoint of the nineteenth century in favor of this strange new world their colleagues have discovered. But one researcher who was far in advance of his fellow biologists was Hans Driesch, who, as mentioned earlier, conceived a vitalist theory of biology following his experiments with sea urchins. Driesch concluded that the development of organisms is directed by, "a unifying non-material mindlike something, . . . an ordering principle which does not add either energy or matter" to the processes it directs.[8] He suggested also that this principle might exist outside the normal framework of time and space—an idea strongly reminiscent of David Bohm's implicate order and the "extracurricular" connectedness of the Einstein–Podolsky–Rosen experiments.

Few of his fellow biologists shared Driesch's view of nature. One exception is Alister Hardy, professor of zoology at Oxford from 1946 to 1963. In 1949 Hardy astonished the British Association for the Advancement of Science by suggesting in his presidential address to the zoological section that telepathy was relevant to biology. In the *Journal* of the Society for Psychical Research, Hardy wrote, "assuming the reality of telepathy, . . . the discovery that individual organisms are somehow in psychical connection across space is, of course, one of the most revolutionary . . . ever made."

Hardy professed himself to be a Darwinist, but it was a strange variety of Darwinism which enabled him to assert that "there is a general subconscious sharing of a form and behavior design, a sort of psychic blueprint between members of a species," and that "the mathematical plans of growth seem to have all the appearance of a pattern outside the physical world which has served as a plan for selective action by way of changing combinations of genes."[9]

The dangerously heretical ideas and experiments of zoologists like Driesch and Hardy were not so much ignored by their fellow biologists as they were mentally quarantined, in case they should prove contagious. If anything, they have proved to be even more infectious than feared.

In the baffling new world of modern physics, scientists find themselves observing and examining a cosmos that has become less and less like a clockwork machine and more like an intelligence. Whether the intelligence is that of ourselves, the observers, or that of the world we examine is not yet clear and perhaps may never become clear. But it would surely be absurd to bestow intelligent characteristics upon the behavior of nuclear particles yet fail to accord such characteristics to living structures.

CONTROVERSIES

CHAPTER 21

The Evolution of Evolution

THE NEO-DARWINIAN IDEA OF EVOLUTION by chance mutation coupled with natural selection has from its inception been welcomed as an extremely powerful tool of explanation. It has traveled far from being used merely to explain physical heredity and the development of biological characteristics. It has been adopted by some of the most distinguished scientific and philosophical minds of the twentieth century to explain phenomena as diverse as animal and human behavior, social movements and trends, and the progressive development of inanimate objects ranging from the elements to the stars, to galaxies and even the universe itself.

This is powerful, heady stuff. But if the idea of neo-Darwinian evolution is unsupported by evidence or experiment when applied to the heredity of plants and animals, what factual basis is there for applying the concept to other natural phenomena?

You don't have to look very far in your local public library to find examples of Darwin's ideas being pressed into service in this or that field. The Dewey decimal catalogue has been almost taken over by Darwinisms: from astronomy to linguistics and from anthropology to law and even religious thinking.

Writing in 1955, Julian Huxley said that

The concept of evolution was soon extended into other than biological fields. Inorganic subjects such as the life histories of stars and formation of the chemical elements on the one hand, and on the other subjects like linguistics, social anthropology, and comparative law and religion, began to be studied from an evolutionary angle, until today we are enabled to see evolution as a universal and all-pervading process.

A little later in the same anthology of science, Huxley goes even further:

Furthermore, with the adoption of the evolutionary approach in non-biological fields, from cosmology to human affairs, we are beginning to realise that biological evolution is only one aspect of evolution in general. Evolution in the extended sense can be defined as a directional and essentially irreversible process occurring in time, which in its course gives rise to an increase of variety and an increasingly high level of organisation in its products. Our present knowledge indeed forces us to the view that the whole of reality *is* evolution—a single process of self transformation.

If true, this would certainly be a fundamental scientific discovery of momentous importance to our understanding of the world. But let's take a moment or two to examine Huxley's definition with the benefit of hindsight. Remember, we are looking for signs of a universally pervasive process that is directional and irreversible, increases variety, and produces higher levels of organization. Is that what we find in nature?

Even a quick glance through the evidence of previous chapters is enough to show that it is not. First, evolution is not directional or irreversible. The kind of primary physical evidence offered for evolution is that of horses, which are always depicted as an unbroken chain of fossils that become progressively larger and more highly adapted through the ages.[1]

The originator of this sequence as a popular illustration, George Simpson of Harvard, asserts that, for instance, the species *Archaeohippus* is a descendant of the ancestral *Mesohippus* from the earlier Oligocene period. Yet the chief characteristic of the more recent *Archaeohippus* is that it is a dwarf or pygmy horse, a major reversal of the previous trend toward steadily increasing size.

This example can be multiplied a hundredfold. Highly ornate extinct shellfish such as ammonites are succeeded in more recent geological strata by simpler and less ornate forms. Many later forms of dinosaur were less ornate in their anatomy than their ancestors.

Turning to the extended meaning of evolution, outside of biology, an often quoted example is the evolution of chemical elements in the nuclear processes in the interior of stars. The energy radiated by stars comes from the fusion of hydrogen atoms into helium, helium into carbon and so on, until heavier and heavier elements such as iron are finally produced. At the end of their lives, many stars detonate in cataclysmic explosions that return these newly formed heavier atoms back to interstellar space where they may later become part of a second and further star systems, in a repetitive process. Some astronomers think it highly probable that a single stellar lifetime is not long enough for substantial amounts of the heaviest elements to be created and several stellar lifetimes are necessary to accumulate the quantities of heavier elements, such as lead and uranium, that we find on the Earth. Thus these elements are said to have evolved. Since our own bodies contain heavier elements such as iron and manganese, then this chemical evolution is an important precursor to biological evolution.

While it is perfectly true that hydrogen atoms are transmuted into heavier and heavier elements in the fusion processes occurring within stars, this process is not irreversible. On the contrary, at the end of their lives many stars explode in a burst of energy that will rip apart a large quantity of heavier atoms, returning them to elementary forms.

Moreover, when the heavier elements that are returned to space condense under gravity to form the nucleus of a new star, some of the heavy elements are pulled apart at high temperatures to form the hydrogen plasma that fuels the stellar fusion process once again.

If evolution is not irreversible, perhaps it leads to greater variety as Huxley claimed? David Raup, professor of paleobiology at the University of Chicago, has made a special study of extinctions. He has pointed out that

Countless species of plants and animals have existed in the history of life on Earth. Estimates of the total progeny of evolution range from 5 to 50 billion species. Yet only an estimated 5 to 50 million species are alive today—a rather poor survival record. With, at the most, only one in every thousand species surviving, what happened to the others?[2]

Far from *increasing* the variety of creatures on Earth, the progress of evolution seems to have had the effect of thinning out the population, and indeed that is the very basis of Darwin's concept—only the fit survive. Evolution in this Darwinian sense can be said to have increased variety if, and only if, you begin with the Darwinian concept of a single or a few organisms as the ancestors of all living things—once again the argument is circular.

Finally, we have the Huxleyan idea that evolution leads to higher levels of organization. Again the real world of natural observations provides plenty of evidence that this idea cannot be correct. A virus is not a more highly organized organism than a self-replicating cell; it is *less* highly organized. Yet viruses must have evolved *after* cells not before, because they can reproduce themselves only by taking over the replication mechanism of a host cell. A snake is not *more* highly organized than a lizard; it is less so because it lacks legs and arms and moves like the primitive worm. Yet Darwinists believe that snakes have evolved from lizardlike creatures—and there are many similar examples of regressive development. If such regression is a natural process in the interests of survival, then why doesn't the whole of nature regress to the genetic immortality of a single-celled organism, which is able to survive the most hostile conditions?

One hundred thirty years after the publication of *The Origin of Species*, Darwinism is still a theory, and still lacking the decisive and incontestable empirical evidence that would end the debate once and for all: that would conclusively demonstrate the correctness of the theory, and ensure its acceptance by the community. Ironically, for most of this century Darwinists have acted as if they had already delivered this conclusive evidence and as though we, the community, had already accepted their theory.

In any other serious scientific discipline, such as physics or chemistry, scientists welcome the opportunity to test a new theory by seeking evidence that would falsify that theory. By contrast, in evolutionary biology, Darwinists avoid evidence that contradicts their theory, while actively seeking and claiming any and all evidence that might tend to support it. For instance, wherever there is any evidence relating to evolution as a principle, Darwinists claim that evidence for their theory of mutation and natural selection.

There is, for example, the very suggestive circumstantial evidence that, since the Eocene, horses have evolved from a small browsing animal with multiple toes to a large grazing animal with a single toe or hoof. (The evidence is fragmentary with no actual chain of proof, but is nevertheless very suggestive.) Darwinists proudly point to the reconstructed lineage of the horse family as evidence for Darwinism. In fact, though, the evolution of horses provides no evidence whatsoever as to mechanism and does not entail evidence for either genetic mutation or natural selection.

In human anthropology, each new fragment of bone or tooth is enthusiastically greeted as further evidence of man's descent by natural selection from an ancestral apelike creature when, as described earlier, every single find of this sort has been definitively assigned to either humans or apes, not to any intermediate category.

This intellectual degeneracy is the outward expression of the fact that neo-Darwinism has ceased to be a scientific theory and has been transformed into an ideology—an overarching belief system that pervades all thinking in the life sciences and beyond.

The replacement of Darwinism-the-scientific-theory by Darwinism-the-ideology has been an important part of twentieth-century political thinking just as it was important to the politics of the nineteenth century. In Darwin's day the theory was accepted partly because it supported the racism and European chauvinism on which the mercantile empire of Britain's ruling class was built and maintained. Today, Darwinism the ideology is one of the principal bulwarks of free-market economic theories and right-wing political thinking. It represents perhaps the most complete absorption of Darwinian thinking outside of the realms of biology.

In a free market, according to economic Darwinists, the factor which guarantees the consumer the lowest prices and highest quality of goods and services is competition. But in any competition there have to be winners and losers (Darwin's struggle for survival). Moreover, there has to be a constant supply of new ideas, new products, and new services to provide the variety on which the natural selection of the marketplace will operate. Thus, in free-market capitalist economies some people *must* fail (companies go under; employees become unemployed) in order for the community to thrive and prosper. The question is, What is the cause of this success and failure?

Darwinists, and supporters of free-market economic policies, say that those who succeed are those who are best fitted or best adapted to the economic environment—in other words the best and the brightest. Those who fail are the weak, the slow, the not so good. This idea is cruel, but it has a certain stark magnificent grandeur about it, a kind of noble savagery. Equally important, it is a perfectly *natural* mechanism. It is merely an extension into human society of the great Darwinian principles of natural selection and the survival of the fittest. Failure in competition may be cruel, but it is fair and just and inevitable, because it is nature's way.

Most important of all, not only is competition a natural process, it is also a healthy one—one that benefits the whole community, in the long run, because it ensures the "evolution" of the most efficient means of producing goods and bringing them to market when and where consumers want them. The human cost of this "evolution" is merely a necessary part of the process and the price that we in Western countries pay for the prosperity we enjoy in comparison with the disastrous performance of the managed economies of Eastern Europe in the recent past.

Many right-wing politicians and economists harbor these ideas in a sort of half-secret way. Because of their innately cruel and antihuman tenor they may not be spoken of directly and aloud except in the sanctity of the political club bar or in the privacy of government office. To speak aloud of these matters would be alarming and frightening to ordinary people, for they smack of Hitler and Nietzsche and ideas of racial purity, and the elimination of specimens that weaken the breed.

Right-wing politicians soften the stark reality of these Darwinian ideas by paying lip service to the need to protect the weak, the ill, the old, and the unfortunate from the ravages of fate. All the while, however, they continue to believe that such "losers" are a necessary part—indeed, an inescapable, essential part—of the economy.

Central to these beliefs and this kind of thinking is the idea that in commerce—as in all things in life— strength, skill, talent, intelligence and bravery are all desirable qualities because they lead on to success in any endeavor. "Fortune favors the brave." "None but the brave deserve the fair." Thus right-wing politicians—most notably in recent years, Margaret Thatcher and Ronald Reagan— were able to equate their political ideas with what they like to describe as the old-fashioned Victorian values.

Like the rest of Darwinism, these ideas seem self-evidently true, until you dig a little deeper. What this thinking disguises is the awkward fact that in commerce, just as in nature, it is impossible to define or test any concept of fitness of purpose because it is impossible to define the fit in any way other than as those who succeed. The fit survive and those who survive are the fit. Just as in evolutionary biology the "survival of the fittest" is no more than a rationalization made retrospectively after the event.

In reality commercial ventures succeed for a whole variety of reasons. Sometimes it is because the entrepreneurs who run the businesses, and the people who work for them, deploy all the desirable Victorian capitalist qualities—hard work; bright ideas; giving the customers what they want. Sometimes it is because the suppliers are protected by a completely artificial and unfair monopoly or near monopoly—like the nuclear power industry, or Bell Telephone before deregulation. Sometimes it is because of a great stroke of good luck—as when the oil companies found huge oil and gas deposits in their backyard.

Failure of businesses can also occur for a variety of reasons. Sometimes, as predicted by the Darwinist model, it is because of laziness, stupidity, bad management, or other failure to compete effectively. But it may also be because legislative changes force costs up, or raw materials become unexpectedly more expensive (perhaps because of war or revolution in some far-away country), or because of some stroke of bad luck—as when disease strikes down the farmer's prize dairy herd.

Politicians are reluctant to accept the implications of this unpleasant fact (just as biologists are). It is that the world is fundamentally chaos-related and its effects on our political and economic
systems are unpredictable. There are just as many entrepreneurs
of intelligence and skill who fail as there are bad managers. And
there are just as many wealthy morons who succeed as there are
hardworking, thrifty, virtuous entrepreneurs. What economic Darwinists do not wish to acknowledge even to themselves is that their
theories are quite incapable of predicting which individuals or which
companies will be the losers and which will be the winners.

This paradox lies at the very heart of a free-market economy, in
its stock markets. If Darwinist theories of economic competition were
true then they would yield reliable predictions and it would be perfectly possible for investors to invest in companies who would always
yield a high rate of return. In reality it remains impossible to obtain
consistently such a high rate of return because the companies that
compose the market are subject to random fluctuations in their for-
tunes which are essentially unpredictable.

Even with centuries of such experience, economic Darwinists
still continue to believe that their theory does predict the outcome
of competition, even though every day of the week some of them
are losing their shirts on the stock exchanges of the world.

It is not only politicians of the right that have espoused Darwinist ideas. Karl Marx was a devout Darwinist and his political
descendants on the left have retained a strongly Darwinist flavor in
their political beliefs. In *Das Kapital*, Marx called Darwin's theory
"epoch making" and said,

> Darwin has interested us in the history of Nature's Tech-
> nology, i.e., in the formation of the organs of plants and
> animals, which organs serve as instruments of production
> and of sustaining life. Does not the history of the produc-
> tive organs of man, of organs that are the material basis of
> all social organisation, deserve equal attention?[3]

In this respect, Marx saw himself as applying the same reductionist
analysis to a material world in which everything from chemistry to
economics to human behavior was ultimately purely mechanical

and could be reduced to its elements through rational analysis. The final social outcome of Marx's thinking has been the planned economies of the former Soviet Union which, unsurprisingly, have turned out also to be chaos-related and incapable of rational management.

Charles and Karl, alike in their ideological domination of much of twentieth-century reductionist thinking, share much the same fate as that century comes to an end.

Darwinists of every stripe (biological, economic, political, and sociological) should celebrate their belief in nineteenth-century values by hanging a Victorian-style embroidered sampler over their beds reading, "The value of shares can go down as well as up." And each night as they say their prayers and climb into bed, they should reflect that no matter how plausible their theory may seem, it is quite incapable of predicting the behavior of anything or anyone.

CHAPTER 22

On Being Thick-Skinned

WHEN CRITICS OF DARWINISM ASSEMBLE their evidence and make their case, it is not unusual for them to torture Darwinists with inexplicable and complex individual examples of structures and behavior from the animal and plant world which defy probability. The ammunition at their disposal is immense and creationists in particular never miss an opportunity to hurl an example or two at evolutionists.

While developing my main arguments, I have resisted the temptation to indulge in this amusing sport, and I have avoided the seductive lure of resting any part of my case on Paley's argument from design—on the improbability of this or that anatomical feature, such as the complexity of the human eye—since I feel that these examples are as likely to cloud the issues as to clarify them. And most of the examples are beginning to become dog-eared from being hurled at Darwinists so often.

But since this book is an attempt to present a global critique of neo-Darwinism, it would be negligent of me to omit an entire body of evidence. So in the interests of completeness (and a little modest entertainment) I present the following golden treasury—or perhaps grimoire—of evolutionary impossibilities.

Darwinists have understandably had to become thick-skinned about such examples being thrown at them. But that is not the meaning of the present chapter heading. Instead it refers to what

some think is the most bafflingly improbable mutation—the thick
skin on the soles of our feet. This thicker skin does not appear after
birth as a result of walking around, but is present in the human
embryo (and the embryos of some other species such as apes). It is
therefore an inherited characteristic. How does it come about that
we have thick skin just where we need it and nowhere else on our
bodies? The Darwinist explanation is that it is the result of a chance
mutation. Presumably other human ancestors had chance mutations that gave them thick skin elsewhere—on their noses perhaps
or their ears, but this did not increase survival chances and hence
was not selected for.

Other species also have thickening of the skin in places
uniquely suited to their mode of life: the African warthog has
callosities on its wrists and forelegs on which it leans while feeding; the camel has them on its knees; that curious bird the ostrich
has them back and front on its underside where it squats. All are
inherited characteristics. All are present just where the animal
needs them and nowhere else. There is no species known which
possesses unnecessary callosities. Does anyone *really* believe this
is the result of random chance?

The human eye is generally taken as the archetypal "impossible" structure. It is the one most often discussed and the one
Darwin himself confessed gave him "a cold shudder." In one sense,
the eye ought not to give evolutionists the shivers because it is only
another structure—admittedly many times more complex than an
arm or a wing, but degree itself is no objection to the principle of
random mutation. Once you have accepted that mutation coupled
with natural selection can produce something as complex as a DNA
molecule or a bacteria, then it is just a matter of time before something as complex as the eye arises. And evolutionists have allotted
themselves practically unlimited time. But it is not the complexity
of the eye itself that causes Darwinists their difficulty. It is the problem of demonstrating all the many stages of the eye in transition.
Consider this statement from Garret Hardin:

> Were all other organisms blind, the animal which managed
> to evolve even a very poor eye would thereby have some
> advantage over others. Oysters have such poor eyes—many

tiny sensitive spots that can do no more than detect changes in the intensity of light. An oyster may not be able to enjoy television, but it can detect a passing shadow, react to it as if it were caused by an approaching predator, and—because it is sometimes right—live another day. By selecting examples from various places in the animal kingdom, we can assemble a nicely graded series of eyes, passing by not too big steps from the primitive eyes of oysters to the excellent (though not perfect) eyes of men and birds. Such a series made up from contemporary species, is not supposed to be the actual historical series; but it shows us how evolution could have occurred.[1]

This view is echoed by Gavin de Beer, an embryologist and director of the British Natural History Museum, in his *Atlas of Evolution* where he illustrates a sequence beginning with the primitive eyespot and culminating in the eye of mammals. "There can be little doubt," he wrote, "that the series of stages described through which the eye passes in embryonic development is a repetition of the manner in which it evolved."[2]

This is a fair summary of the Darwinist view. But the difficulty with Hardin's argument is that it specifically fails to do what he sets out to do—to demonstrate step-by-step the evolution of the human eye. It says, in effect, that all the species in the living world today have evolved by random mutation and natural selection: they exhibit various kinds of eye from primitive to advanced; therefore the human eye has developed by such evolutionary stages. Hardin has reached his conclusion only by including it in his premises.

The fact that an oyster has a primitive eye does not demonstrate that complex eyes evolve from primitive eyes—that is the very matter in question. If paleontologists could produce a series of fossil mammals, or reptiles, or fish showing the eye in these various stages, their case would be made. But of course, if they could produce such a series of fossils, they would not need to concern themselves with medieval debates about eye complexity—they would already have made their case.

Interestingly, Professor Wolsky in his book *The Mechanism of Evolution* points out that light-sensitive organs in all creatures seem to have evolved in the places where light falls most intensely, suggesting that—just like thick skin on the feet—this appears to be some form of design.[3] The Darwinists' traditional response is that mutations that cause eyes in the "wrong" place would not be adaptive and hence would not be selected for. But here they are attempting to have their cake and eat it, for they also argue that (in Hardin's words) "even a very poor eye would have some advantage." Hence we should expect to find creatures with eyes in less-than-optimum locations, such as on the flanks or the base of the spine. But no such creatures exist, either today or as fossils.

My next example is not so much concerned with the evolution of new organs as the disappearance of existing ones. Evolutionists believe that, for example, the snake is a reptile which was originally like a lizard, but has lost its arms and legs as a result of adapting to a crawling mode of life. Similarly, the whale is believed by evolutionists to be a mammal which has returned to the sea, and lost its limbs in order to become streamlined for swimming. Despite the whale's enormous size, its thighbone has now shrunk to a mere eighteen inches long and is on its way to vanishing entirely.

The question is, What was the evolutionary advantage of the thighbone becoming any smaller than the whale's streamlined body envelope? What was the evolutionary advantage of the snake's arms and legs disappearing altogether? Or the mole's eye sockets being filled with muscle? Is it really rational to suppose that *random* mutations appeared which progressively diminished just these organs until they vanished entirely, long after any survival advantage could have been gained? The concepts of mutation and selection are both flawed in explaining the whole field of regressive organs. It seems clear that some systematic process or program is taking place which, once initiated, proceeds to a conclusion. Where does the "program" reside? How does the "system" know when to start and stop?

One category of impossible mutations has to do with precision engineering: engineering to limits that we would find extremely difficult to emulate. The often quoted eye is in fact not very precisely engineered: its elements can vary by a substantial margin and the eye will still function reasonably well. Some natural structures, though, require an accuracy of millionths of a centimeter. The silvery skin of fish is designed to provide a reflective surface that enables them to remain camouflaged and unnoticed by predators, in the greenish gloom of the sea. To achieve this, fish secrete millions of tiny nitrogenous crystals in layers on their skin and scales. But this is not all. To increase the efficiency of their reflective coating (from about 25 percent reflective to as much as 75 percent) the fish secrete multiple layers of mirror crystals sandwiched between layers of cell tissue. But to be effective, the "sandwich" has to be an exact thickness—exactly one-quarter of the wavelength of the incident light. For the greenish light of the undersea world, this means a separation of seven millionths of a centimeter.[4] Does anyone really believe that this precision was achieved by random mutation?

An important area of biology in which the neo-Darwinist theory is an inadequate tool of explanation, and one that leaves a disturbingly large blank on the scientific map, is that of behavior. There is ample evidence that the young of many species are born with highly specialized abilities that they cannot learn from their parents or others of their species and which therefore must be inherited.

One of the most striking examples of this kind of behavior is that of the cuckoo. As is well known, the hen bird lays her egg in the nest of another species. The cuckoo's parents both migrate some 12,000 miles to South Africa while the cuckoo chick hatches and attempts to tip his rival chicks out of the nest.

Once the young cuckoo is fledged and grown it, too, will fly 12,000 miles south to join the parents it has never met at the winter quarters it has never seen, with perfect navigational accuracy.

The only mechanism that exists in the neo-Darwinian theory to account for this complex behavior is Mendelian genetics—the belief that there is a gene for navigating 12,000 miles to an unknown place. This is the sort of proposition which—if put forward today—would attract the skeptical response, "extraordinary claims require extraordinary evidence." So far, no evidence at all, ordinary or extraordinary, has been put forward in support of this idea.

My personal favorite among the specimens in the black museum of incredible mutations, is the general matter of the alternation of generations. This is seen for instance in jellyfish who reproduce by releasing eggs and sperm into the sea. The fertilized egg does not develop into another jellyfish straightaway but settles down to another form of life as a flowerlike polyp anchored to the sea bottom or a rock. Ultimately the polyp buds (in a different way from its parent jellyfish) and the buds grow into free-swimming jellyfish once more. In some types, the majority of time is spent in the free-swimming form with only relatively short spells as a polyp. The common sea anemone, on the other hand, spends most of its time anchored to rocks and little as a free swimmer.

The alternation of generations raises all kinds of fascinating questions concerning the adaptive advantage of such a way of life, and how it could have come about by microscopic mutations. It is hard to imagine how the alternation of generations could come about a little at a time—indeed this is one of the examples that made Richard Goldschmidt conceive his hopeful monster theory.

The aspect that fascinates me most is that there is some kind of counting or timing mechanism at work here: a mechanism that recurs in animal and plant life. A few examples will explain. The artichoke plant, grown by gardeners for its fruit, will crop for three years; the plant then dies, or sometimes lives on but will crop no more. However, if a cutting is taken and planted, it too will crop for three years. The common variety of asparagus crown will crop for seventeen years and then cease. Human children have two sets of teeth: the first set come through in a miniature size appropriate to a child; the second set of teeth come through full grown at adult size even though they usually appear when a child is around only seven years old. There is a species of bamboo tree that flowers every 117 years, and cacti that flower every twelve years. The ptarmigan and the Arctic fox assume a whitish coat in winter and a brownish one in summer.

How does the sea anemone "know" it is time to become a jellyfish? How does the artichoke "know" its three years are up? How do the child's teeth "know" they are second teeth and must be bigger than the first? How do they know what scale to be on at all? How do the ptarmigan and Arctic fox "know" when to change coat? And when to change back again?

The answer may be a relatively simple matter of genetic coding. For example, adult-sized teeth may be the product of a genetically coded scale factor that is applied to every protein synthesis regardless of its function in the body. But it is very hard to see how a timing mechanism can operate—especially across the generations—without some global or systematic function being invoked that controls the entire organism in some way. And this is specifically what Darwinists say does not exist.

In 1940 Richard Goldschmidt felt concerned enough about the conventional neo-Darwinist view to throw down this challenge:

> I may challenge the adherents of the strictly Darwinian view
> . . . to try to explain the evolution of the following features
> by accumulation and selection of small mutants: hair in
> mammals, feathers in birds, segmentation of arthropods and
> vertebrates, the transformation of the gill arches in phy-
> logeny including the aortic arches, muscles, nerves, etc.;
> further, teeth, shells of molluscs, ectoskeletons, compound
> eyes, blood circulation, alternation of generations, stato-
> cysts, ambulacral system of echinoderms, pedicellaria of the
> same, cnidocysts, poison apparatus of snakes, whalebone,
> and finally primary chemical differences like haemoglobin
> versus haemocyanin, etc.

Goldschmidt adds that corresponding examples from the plant world could also be given.[5]

So far as I am aware, no Darwinist has accepted Goldschmidt's challenge. But whereas he was regarded as having a screw loose in 1940, he is taken a great deal more seriously today.

CHAPTER 23

The Fish That Walked

THE WRITER G. K. CHESTERTON tells us that

> God made the wicked grocer
> For a mystery and a sign
> That men might shun the awful shops
> And go to inns to dine.

If God exists and if, as Chesterton thought, he possessed a sense of humor when creating the world, he must surely have created that mysterious and extraordinary creature the coelacanth to provide mankind with a little light entertainment on wet Sunday afternoons. The story of the coelacanth is worth recounting if only because it reminds us how easy it is for science to get things wrong.

Like most human affairs, science is prone to extraordinary co-incidences. On Saint Valentine's Day in 1876 for example, two men walked into the U.S. patent office, each with the same invention under his arm. Alexander Graham Bell and Elisha Gray both filed patents for the telephone on that February day, giving rise to a protracted lawsuit over who had priority—an honor that the courts, and the history books, have awarded to Bell.

The dust had hardly settled on that lawsuit when, a decade later in 1886, Charles Hall in the United States and Paul Héroult in France simultaneously but independently devised the electrolytic method for producing aluminium on a commercial scale. It is easy to dismiss these coincidences on the grounds that people working in similar technical fields are likely to come up with similar results: that the coincidences are, so to speak, "rational coincidences."

There is, though, another kind of coincidence—the kind of wholly irrational and unpredictable event which Carl Jung termed "synchronistic" and which has an almost mystical quality. Two of the inventors mentioned earlier, the American, Hall, and the Frenchman, Héroult, as well as making the same discovery in the same year 1886, were both born in 1863 and both died in 1914—a coincidence which reason is powerless to explain.

For those who, like me, are collectors of coincidences, the Darwinian theory of evolution is a veritable gold mine of improbable events. Few of these incidents have proved quite so extraordinary as a discovery made by fishermen off the coast of Africa in 1938. This discovery resulted in the resurrection of a long-dead witness for the prosecution against Darwin—the ghost of the fish that walked.

To appreciate the full significance of the fishermen's strange haul, it is first necessary to go back almost exactly a century earlier, to the survey vessel *HMS Beagle* and to the young Charles Darwin returning home from his five-year voyage of natural history discovery in 1836.

On board the *Beagle*, surrounded by fossil remains from distant continents, Darwin began to contemplate the idea of evolution from simple organisms to more complex ones, under the hidden hand of natural selection. The difficulty this has led Darwinists into, as we have seen, is the failure to find any transitional species in the fossil record—or as the newspapers were later to dub them, the "missing links" in the chain of life.

The missing links looked for were not merely human, but included every part of the animal kingdom: from whelks to whales and from bacteria to bactrian camels. Darwin and his successors envisaged a process that would begin with simple marine organisms living in ancient seas, progressing through fishes, to amphibians—living partly in the sea and partly on land—and hence on to reptiles, mammals, and eventually the primates, including humans.

But although each of these classes is well represented in the fossil record, as of yet no one has discovered a fossil creature that is indisputably transitional between one species and another species. Not a single undisputed "missing link" has been found in all the exposed rocks of the Earth's crust despite the most careful and extensive searches.

This is a difficulty because, if life has evolved in the way that Darwin proposed, there should be many millions of transitional species—invertebrates with rudimentary backbones; fish with incipient legs; reptiles with half-formed wings, and so on. Indeed, given a theory that postulates continuous random genetic mutation, and hence a continuous spectrum of life forms, constantly evolving to become better and better adapted, such specimens should be the rule rather than the exception. Life itself should be boldly innovative, rather than cautiously conservative.

At first the lack of missing links could be attributed to the fact that much of the world remained unexplored. Darwin himself expressed the hope that further exploration would turn up the missing fossils. But the hope gradually faded until it became clear that paleontology had accumulated an almost unmanageably rich collection of specimens but the fossil record nevertheless continued to be comprised mainly of gaps.

By the time the First World War had ended and the new century was under way it had become abundantly clear that earlier hopes of finding fossils to fill in the many gaps were wearing rather thin. Further exploration and collecting were merely adding more of the same sort of fossils that were already known and catalogued. Museum departments therefore turned their attention to making sense out of the millions of specimens they already had in their glass cabinets and store rooms.

These researchers naturally looked to comparative anatomy as their guide and focused much of their attention on the major question of the transition from the era of exclusively marine life, to that of life on the land. They correctly foresaw that if they could provide detailed evidence of this transition—the first and most important of all—they would provide powerful evidence in favor of the Darwinist model.

Much debate ensued in the paleontology departments of the world's natural history museums as anatomists examined and rejected, one after another, candidates for the progenitor of all ter-restrial life: the fish that had, after millions of years of life in the sea, finally crawled and flapped, gasping, onto the mud of some ancient estuary to lay its eggs.

The material they had to choose from was vast. Of all fossils, those of marine creatures are by far the most plentiful because of their greater populations compared with terrestrial animals and because of the more favorable conditions of preservation in ocean sediments. But certain fundamental requirements were logically obvious from the start. The candidate would be found among the "bony" fishes rather than among those with merely a flexible carti-laginous skeleton. It must have a well-developed bony skull. And, most important of all, it must have four fleshy fins, supported on bony growths, to enable its colonization of the land, and from which the four-limbed pattern of life could have evolved.

These requirements narrowed the field considerably, and with a rare unanimity paleontologists agreed that they had found their fish. At last the cases of the various claimants had been examined and the impostors rejected; the pedigree and credentials of the suc-cessful candidate were prepared, and he was spruced up for presen-tation to his waiting public. The "press was squared, the middle classes all prepared," as Hilaire Belloc observed of a young hopeful in somewhat similar circumstances.*

The fish that had walked, it was announced, was of the Cros-sopterygian (or bony-skulled) class, and more specifically was a Rhipidistian (or lungfish). The fish in question had been extinct for a long time, along with all its close relatives. But its anatomy was well known from the hundreds of specimens found through-out the fossil record in many parts of the world, right up to its extinction at about the same time as the dinosaurs died out, in the Cretaceous period.

One particular example of the ancestral fish gave paleontolo-gists abundant fossil material to study—a fish of the genus *Coelacanthus*. Coelacanths had been found in places as far apart as New Jersey, Greenland, Bavaria, Spitzbergen, Brazil and at several

---

* Lord Lundy, who was "destined to be, the next Prime Minister but three."

places in Britain. The coelacanth had been described by the pioneer paleontologist Gideon Mantell in the early nineteenth century and had been illustrated by Darwin's champion, Thomas Huxley, in 1866.

Specimens of the fish had been preserved in fine detail in ancient rocks and its anatomy had been well studied and catalogued. And it was its anatomical features—plus a little intelligent guesswork—that prompted such unanimity among its authors. The fish and its relatives had flourished during the Devonian period some 350 million years ago, before declining to a dignified end. But before expiring, it had managed to flap onto the estuarine mudflats with the aid of its embryonic limbs, and give birth to a hopeful new generation of creatures who were able to exploit the land—truly a Columbus among marine organisms and a worthy progenitor of the human race.

The announcement of the discovery of the "missing link" was one of Fleet Street's earliest scientific scoops. And although the readers of the popular dailies couldn't tell a coelacanth from a breakfast kipper, the public imagination was fired by the discovery. The British Museum of Natural History mounted a display and parties of schoolchildren, in pursuit of merit marks from approving schoolteachers, pressed their noses against the glass cabinets of South Kensington.

Those responsible for filling the glass cabinets, and the minds behind the noses pressed against them, probably permitted themselves a moment of self-congratulation. If so, it was short-lived. For at precisely that moment, the most astonishing and irrational coincidence occurred.

Fishermen trawling the waters off East London on the coast of Africa in 1938 found a strange-looking fish in their nets. The decomposing—and by now highly aromatic—remains of the fish were examined by the curator of the East London Museum, Margaret Courtenay-Latimer, and by Professor J. C. B. Smith of Rhodes University, South Africa, who identified it as a living specimen of the coelacanth.[1]

The strange catch was a "living fossil" and its discovery must have been poetically inspired by the goddess of coincidence to remind mortals of their fallibility.

It soon became clear from examining the strange catch that the coelacanth was a poor choice for the "missing link" between marine and terrestrial life. Its four fins are much like those of any other fish and are no more suitable for supporting its weight on land, or of giving rise to amphibious limbs, than those of a fairground goldfish. There is, too, the awkward fact that the coelacanth lives at such great depths in the ocean (up to 200 meters) that it explodes due to decompression when brought up to the surface—a slightly ticklish handicap for a colonizer of the land. In 1986 Hans Fricke of the Max Planck Institute for Animal Behavior used underwater video cameras to observed the coelacanth in its natural habitat. Unsurprisingly, the coelacanth does not stroll on the seabed with its fins, as supposed, but swims through the water just like any other fish.

Back in the bone departments, the innocent coelacanth was stripped of its title and dignities in a purely private ceremony. The official line today is that the coelacanth was merely an evolutionary dead end and some other creature—possibly *Eusthenopteron*—holds the coveted "missing link" title. *Eusthenopteron*, too, is supposed to be extinct—let us keep our fingers crossed and hope that, this time, it stays dead.

This chapter, no more than a piece of fun, might be subtitled "a cautionary tale." Its story holds a number of lessons both for those who choose to believe in the synthetic or neo-Darwinist theory of evolution and, also, for those who do not believe in it.

The tale of the "fish that walked" is a cautionary tale in more ways than one. It cautions us against blind acceptance of the intellectual appeal of an elegant theory, and against uncritical acceptance of the intellectual authority of those whom we, as a community, pay to do our difficult thinking. Scientists are today's Magi or wise men. One of their main functions is to satisfy public curiosity about natural events. But being only human, scientists are sometimes driven to their conclusions by the weight of public demands for knowledge, rather than led to them by the weight of evidence.

When, for instance, J. J. Thomson discovered the electron in 1897, great public and academic interest was aroused and thereafter Thomson was besieged with demands from students and members of the public wanting to know, "What is an atom like?" Under such pressure, Thomson hazarded the speculation that an atom resembles an apple with the electrons embedded inside the nucleus like pips, an idea now known to be false. No one would blame the scientist for entertaining a hypothesis that later proves to be false; indeed that is how science proceeds. But in the case of atomic science, the subject matter is always present before us for further investigation. The tracks of atomic particles are visible to all in the cloud chamber and errors of theory may be corrected by further observation.

Questions concerning the origin of life, though, are a different matter. Past biological events are no longer available for observation and, regrettably, the tracks they have left are obscure. The traces of biological history that do remain present a vast and often puzzling picture, a picture that can be grasped only by the construction of suitable models. The neo-Darwinist theory is perhaps the most elegant and powerful model ever constructed in the life sciences. But like all models of the real world, it has ultimately reached a point where it is no longer able to contain the data it seeks to explain.

CHAPTER 24

Angels Versus Apes

D IRECT CONFLICT AND CONFRONTATION with religious belief was built into Darwinism from the outset. Darwin expected trouble and, being a retiring sort, did not relish the prospect. But his great champion, Thomas Huxley, certainly did savor the cut and thrust of scientific debate. And after urging Darwin to make his findings public, Huxley confided to Darwin on the eve of publication that he was "sharpening up my claws and beak in readiness."

In less than a year, Huxley found the major opportunity he sought for public debate at the notorious British Association meeting in Oxford in June 1860. The debate over Darwin's newly published theory took place in the library of the university's museum at the end of a week of meetings where the explosive issue was never far below the surface, threatening to ignite at any time. By Saturday, tension was high and some 700 people, including groups of cheering and counter-cheering students, crammed noisily into the library, forcing the speakers to shout to make themselves heard above the din.

The combatants this cheerful mob had come to hear were Huxley, championing Darwin, and Samuel Wilberforce, bishop of Oxford and fellow of All Souls, a scintillating orator who represented the theological faction. According to Charles Lyell, who was present, Wilberforce began well, launching into a series of calculated and savage attacks which drew loud applause from his

supporters. Having warmed up his audience, however, Wilberforce made the tactical error of launching a personal attack on
Huxley.

Turning to the young geologist, Wilberforce asked him whether
he was related to the apes on his grandfather's side or his grand-
mother's.

Huxley replied to the bishop's scientific arguments with "force
and eloquence." He then addressed the personal remark and told
Wilberforce:

> A man has no reason to be ashamed of having an ape for a
> grandfather or grandmother. If I had a choice of ancestor,
> whether it should be an ape, or one who having a scholastic
> education should use his logic to mislead an untutored pub-
> lic, and should treat not with argument but with ridicule
> the facts and reasoning adduced in support of a grave and
> serious philosophical question, I would not hesitate for a
> moment to prefer the ape.

Huxley was judged by Lyell to have got the better of the debate on
this occasion but, then as now, nothing is ever settled merely by
debating the question of creation versus evolution. The two sides
are as deeply entrenched in the 1990s as they were in the 1860s.
Attacks by religious believers remained confined to the debating
chamber for some decades after Huxley confronted the Bishop.
But in the early twentieth century Darwinism began to be taught
in schools and this gave religious groups a battleground on which
to fight. The first result was the famous Scopes trial in Tennessee
in 1925.

In March 1925 Bible fundamentalists in Tennessee instigated
the passing by the state legislature of a law forbidding the teaching
of any doctrine denying the creation of humans as taught by the
Bible. The American Civil Liberties Union decided to contest this
law and a young school teacher, John Scopes of Dayton, volun-
teered himself as a defendant. The trial became a confrontation
not only of fundamentalists versus evolutionists but also of two
great public figures, William Jennings Bryan, the prosecuting attorney, and Clarence Darrow for the defense. Although he pros-

ecuted this case, Bryan was a lifelong champion of liberal causes. Darrow was an advocate of freedom of expression and a highly successful criminal lawyer with an appetite for defending a mighty cause.

The trial was a disappointment to both sides because the judge ruled that the issue of Darwinism itself was not to be tried. The trial was to confine itself solely to the question of whether Scopes had broken state law, which was not disputed. Scopes was found guilty and fined $100, although the conviction was later overturned by the supreme court on the technical grounds that the penalty exacted was beyond the powers of the court to impose. The law under which he was prosecuted remained on the statute books of Tennessee until 1967.

Although the proceedings of the Scopes trial contributed nothing concrete to the debate, they do provide some further insight into the spread of Darwinist ideas into mainstream education. At the trial some interesting exhibits were introduced as evidence in order to establish a factual basis for the teaching of evolution. These exhibits included our old friend "Piltdown man" together with a tooth which was the sole fossil remains of his American counterpart *Hesperopithecus*—"western ape man."

The Piltdown finds had not yet been recognized as the creation of a practical joker who had cleverly planted an orangutan's jaw with a human skull in a gravel pit. The remains of the "earliest Englishman" were displayed to the hushed courtroom as positive proof of humankind's simian ancestry. The other fossil was a tooth that had been found by amateur geologist Harold Cook, in 1922 in Pliocene deposits in Nebraska. Cook sent the tooth to Henry Fairfield Osborn, eminent Director of the American Museum of Natural History. Osborn believed he could see anatomical features of both ape and man in this tooth and that it proved man had descended from apes in America as well as Europe and Asia. The new world could now lay claim to a little paleontological glory just like the old.

Some years after the trial, an expedition from the American Museum of Natural History returned to the place where Cook had made his discovery and excavated a number of similar teeth. These showed that *Hesperopithecus* was not a man but an extinct peccary or pig.

So, in so far as evidence was produced to provide a factual basis for teaching evolution, that evidence was actually entirely bogus. This resulted in one case from a deliberate attempt to deceive (though not that of the defense) and in the other from a piece of overenthusiastic identification by the American Museum of Natural History, which was keen to keep up with the paleontological Jones's.

Although the trial itself accomplished little, later in the century the nature of fundamentalist objection gradually changed and became far more sophisticated. In 1964, for instance, religious leaders in Texas objected to the State Board of Education approving biology textbooks containing Darwin's theory. On this occasion the objectors were overruled, but in 1969 similar objections in California were successful when the State Board of Education decided that in the future textbooks should present Darwinism as merely one of many competing theories. In support of their case, the protesters quoted Mayr as saying, "The basic theory is in many instances hardly more than a postulate and its application raises numerous questions in almost every concrete case."

The greater sophistication of religious objectors today consists in their not just opposing Darwinism as blasphemous but in arguing that Darwinism is merely one theory among many and that it should not be taught as the sole repository of truth.

In the 1950s creationists and religious fundamentalists were not taken seriously by science and were regarded largely as a nuisance, while scientific criticism of Darwinism was regarded as unthinkable. Few people in the academic community took any notice in 1957 when Melvin Cook's paper about atmospheric helium was published in the columns of *Nature* magazine. Although in the 1960s, says Cook, a more detailed manuscript, "not unexpectedly nor without some cause, met with considerable opposition and was not published."

Publication of Cook's *Prehistory and Earth Models* in 1966 gave a boost to the embryonic creation science movement, a group of religious believers many of whose adherents are also professional scientists. This group has proved to be quite a thorn in the side of proponents of the synthetic theory, because their expertise has been employed to turn the tables on Darwinists by applying scientific methods. The strategy of this group has been not to emphasize the extent to which Darwinism contradicts the religious teaching of the Bible, but the extent to which it is contradicted by other scientific evidence.

A number of such groups flourished in the United States in the early 1970s, perhaps the best known of which is the Institute for Creation Research in California, directed by Dr. Henry Morris and Dr. Duane Gish. The early products of these creation science organizations were books and magazines that made some impact, especially on undergraduates and younger people. Instead of dogmatic bible quotations and threats of hellfire for atheists, the new generation of creation science publications were often academically researched texts, giving scholarly references, usually to peer-reviewed professional journals.

The early success of scientific creationists showed that they had struck a chord with many people in America, where Darwinism has always been deeply distrusted, and this in turn had the effect of putting Darwinists on their guard against a new kind of threat to their scientific authority. From now on, anyone attacking Darwinism—whether from a religious or scientific viewpoint—would receive a calculated response: condescension and ridicule if their objections were ill-informed; fierce concerted opposition if they contained scientific merit.

Today it would be virtually impossible for any scientific paper that has anti-Darwinian implications to be published in *Nature* or in any serious peer-reviewed scientific journal, regardless of the scientific merits of its findings. To be an exception to this rule an anti-Darwinian paper would have to be of paradigm-shattering importance, like Guy Berthault's papers on sedimentation or Cairns and Hall's experiment on directed mutation. Even then, publication of the results is likely to be hedged around with qualifications, *argumenta ad hominem* directed at the authors and technical quibbles that would never be directed at any paper supporting Darwinism.

A prime example of this academic censorship is the case of British biologist Warwick Collins. In 1976 Collins was studying biology at Sussex University under the eminent Darwinist Professor John Maynard Smith. Collins wrote a paper on sexual selection as an anomaly in Darwinian theory. Dr. John Thoday, professor of genetics at Cambridge, invited Collins to present an expanded version of his paper to an international conference of population geneticists—an honor for the young undergraduate.

Collins says,

> In the paper I tried to extend further my doubts about the assumptions in Darwinian evolutionary theory. Out of courtesy I circulated the expanded paper to my distinguished tutor prior to the conference. Before I was due to take the stand, Professor Maynard Smith stood up in front of the conference and roundly denounced the premises of my paper.

After the conference Maynard Smith told Collins that "he would use his considerable influence to block publication of any further papers of [Collins's] which questioned the fundamental premises of Darwinian theory."[1]

Collins has, indeed, found it impossible to have any further papers published up to as recently as 1994, when a paper he submitted to *Nature* was rejected without reason. Not surprisingly, Collins has left the field of biology.

Darwinists have thus begun not merely to react to criticism by members of their own profession, and by creationists, but have gone on the attack. As in the case above, some of their methods of attack leave a very unpleasant taste in the mouth of anyone educated in the Western liberal-intellectual tradition.

In 1980 the conflict between Darwinists and creationists was escalated even further by Kelly Seagraves, director of the Creation-Science Research Center in California. Seagraves brought a civil case against the state, alleging that, by teaching Darwinism as fact in the science classrooms of the public schools, the state was violating the constitutional rights of his three sons, Kasey, Jason, and Kevin.

One result of this challenge was that the deputy attorney general for the state of California, Robert Tyler, assembled a team of Darwinist scientists willing to defend the State's teaching in court, an echo of the Scopes trial from more than fifty years before. The team included many distinguished American scientists and well-known names including Francisco Ayala of the University of California at Davis, Harvard's Stephen Jay Gould, and Carl Sagan of Cornell University. The team also included G. Brent Dalrymple, a research geologist from the U.S. Geological Survey experienced in radiometric dating.

When the trial took place, the scientific team was not called to the stand and Seagraves lost his case. But the incident had later consequences. Dalrymple, as a result of his interest in the case, started writing papers in rebuttal of creationist geological arguments. In 1991, he published a book through Stanford University Press called *The Age of the Earth* which is primarily a defense of radiometric dating techniques against scientific critics such as Melvin Cook.

In his book and in several papers[2] Dalrymple sets out vigorously to explode what he sees as false creationist arguments and objections to radiometric dating. His writings have become almost a battle flag around which Darwinist forces have rallied in recent years. Whenever a critic of Darwinism raises objections relating to the geological history of the Earth (especially on the Internet) Darwinists now invoke Dalrymple's name like a talisman. If Dalrymple says a certain objection is wrong and has been debunked, then there is nothing more to be said on the matter—except by brain-dead fundamentalists. In fact, as indicated in Chapter 5, Dalrymple's writings are often strong on rhetoric but weak on scientific fact.

One result of these concerted efforts by Darwinist vigilantes to head off or suppress any dissent is that the subject of Darwinism has largely disappeared from the agenda of public debate, both in scientific journals and the popular press.

Dr. Jerry Bergman, professor of biology at Northwest College, Ohio, has made a study of the censorship of papers from scientists who are also creationists. Writing in *Creation Ex Nihilo Technical Journal*, Bergman says,

> If authors are known as creationists, their articles, regardless of the empirical merit and quality, are most often rejected for publication. At times they are accepted, but when the creationist persuasion of the authors is discovered, they are not uncommonly rescinded.

Even articles discussing censorship of creationism are often censored from journals which deal with library censorship. Some creationists find far more success when they publish under a pseudonym or stay in the closet about their creationism. Censorship because of the philosophical and religious orientation of the writer is clearly bigotry.[3]

It is impossible to disagree with Dr. Bergman's conclusion about this kind of behavior. Anyone who doubts that such bigotry exists should consider the case of science journalist Forrest Mims. In 1991, Mims was asked by *Scientific American* to take over its most popular column, "The Amateur Scientist."

Mims says, "During the course of a meeting with Jonathan Piel, the editor, in New York, I happened to mention that I write for a variety of magazines, including Christian magazines. Piel then asked what kind of Christian magazines. I stated I had written a few articles on how to take church kids on long distance bicycle trips. Piel, obviously agitated, then asked, 'Do you believe in Darwinian evolution?' Knowing the consequences, I responded, 'No, and neither does Stephen Jay Gould.'"

A few months later, Piel cancelled Mims's assignment to write the column, because he feared the magazine would be embarrassed should Mims's beliefs become known.

"I did publish three columns in the magazine," says Mims, "but only after the magazine's president intervened. I did not sue the magazine. Their lawyers did, however, send me various threatening communications in an effort to keep me from speaking out on the matter."

Scientific American now has a new editor and things are looking brighter for Mims. The magazine has published two of his letters and is now reviewing an article. Mims has also been invited to make further submissions, although he no longer writes "The Amateur Scientist" column.

Mims also wrote a letter of complaint to the American Association for the Advancement of Science. The Association's committee which considers human rights abuses accepted his letter of complaint and voted sixteen-to-zero to endorse Mims's right to hold his own religious views.

Mims told me, "Good science requires skepticism. Many of us who are skeptical of Darwinism are concerned that philosophical agendas have interfered with and even blocked solid science."

The taboo on debating Darwinism extends to the broadcast media as well. In general, U.S. television networks avoid taking any serious stand on controversial scientific subjects like Darwinism. They know that if they do they can expect to be barraged with concerted complaints, and demands for retraction and suppression of the offending film by a number of voluble academics. Not surprisingly, few producers and directors are willing to run the gauntlet of such treatment.

One rare and honorable exception was NBC's decision in 1996 to broadcast the film *The Mysterious Origins of Man*, made by Emmy-award winning director Bill Cote, in which I and other independent investigators had a rare opportunity to present anomalous evidence of historical geology, and man's past, so that viewers could evaluate this alternative evidence for themselves.[4]

The program proved immensely popular with many viewers, attracting audiences of around 20 million on each of the two occasions when it was shown. The producers also received dozens of abusive responses, which included virtually no attempts to rebut the scientific issues raised but took the consensus position that students and the public should not be given access to such contradictory evidence. They included terms such as; "horrible," "atrocious," "garbage," "anti-intellectual trash," "evil," "deliberate, fraudulent misinformation," "claptrap," "utter rubbish," "nonsense," "unadulterated hogwash," "bullshit," "a piece of junk," "crap," and "shame on you, liars and opportunists."

You might imagine that these remarks came from the keyboards of pharmaceutically challenged undergraduates or semiliterate teenagers. In fact they are the words of senior scientists and academics (including several professors) from the University of California at Berkeley, State University of New York, and Wisconsin, New Mexico State, Colorado, Northwestern, and other universities.

It is unlikely that any film such as *The Mysterious Origins of Man* would ever be shown in Britain, where Darwinism is such a strictly observed taboo subject that no science program has ever been shown or is ever likely to be shown questioning any aspect of the Darwinian theory. One British filmmaker told me that few of his fellow directors would risk making a television film that questioned Darwinism because to question such a scientific sacred cow would be bad for his or her career.

Curiously neither the press nor television in America or Britain feels any compunction about airing highly contentious political or social issues. As a matter of fact they will risk considerable controversy to assert their right to cover what they consider to be in the public interest, thus properly fulfilling their role as champions of the public's right to know about things done with its money and in its name. But when it comes to contentious scientific matters, they become much more reticent. This is probably because if they dare to give space or air time to political controversy, they are merely branded as troublemakers, which is good for their image; whereas if they give such attention to taboo science subjects, they risk being derided as crackpots.

I experienced this kind of witch-hunting activity by the Darwinist police when I first published *Shattering the Myths of Darwinism* and found myself subjected to a campaign of vilification. I had expected controversy and heated debate, because that is in the nature of Darwinism. But it was deeply disappointing to find myself being described by a prominent academic, Oxford zoologist Richard Dawkins, as "loony," "stupid," and "in need of psychiatric help" in response to purely scientific reporting.

It was equally unpleasant to discover that, behind my back, Dawkins was writing letters to newspaper editors alleging that I am a secret creationist and hence not to be believed. This kind of behavior culminated in March 1995 when a British weekly newspaper, the *Times Higher Educational Supplement* commissioned me as a freelance journalist to write a critique of Darwinism and trailered the article in one of its editions by saying, "Next Week: Darwinism—Richard Milton goes on the attack." Dawkins contacted the editor, Auriol Stevens, falsely alleged that I am a secret creationist, and covertly lobbied against the publication of my article, although he had not seen it. As a former newspaper editor myself, I am ashamed to say that the editor of the paper gave in to this bullying and suppressed my article.

The attempted censorship failed because I published the article widely on the Internet, putting it into the public domain and making many in the academic world aware of the extreme lengths to which some of their colleagues are prepared to go to censor free discussion. Not long after, in 1996, an American geologist, David Leveson of New York University, attacked me in the *Journal of Geoscience Education* alleging that I am a "creationist ally."[5]

I found this kind of bullying, bad faith, and intellectual dishonesty in prominent academics both depressing and a little disturbing. It is like lifting a corner of the veil of civilized behavior and finding something very much like intellectual fascism hiding underneath. Most liberal-minded people who have not themselves experienced this kind of thing will find it hard to believe such behavior takes place in civilized society, since there is little sign of it on the surface unless, like me, you begin to ask controversial questions.

Let me make it unambiguously clear that I am not a creationist, nor do I have any religious beliefs of any kind. I am a professional writer and journalist who specializes in writing about science and technology and who writes about matters that I believe are of public interest.

For anyone, anywhere, to say that I am a creationist, a secret creationist, a "creationist ally," or any other such weasel-word formulation, is an act of intellectual dishonesty by those who have no other answer to the scientific objections I have raised publicly.

Most scientists privately accept that there are serious objections to Darwinism such as those cited in this book and privately they will admit to the objections. However, they have become reluctant to discuss them in public (and in a forum like the Internet they will deny them altogether) because they fear that they will aid their critics and unwittingly discredit their own profession. In some cases, they feel it is better to be discreet, pretend that there is nothing wrong, or even to tell a "little white lie" in the interests of the greater good of science.

But despite this closing of the ranks to silence public debate, Darwinism still has a large number of critics and it isn't only creationists who have serious doubts about the theory or who have questioned the established view of historical geology. One legacy of the research and writing of Immanuel Velikovsky, referred to in Chapter 9, is the Society for Interdisciplinary Studies (SIS) which is a forum for scientific discussion of geological catastrophe theories. An organization such as the SIS tends to be written off by Darwinists as a club for crackpots. Yet the list of guest speakers at the July 1997 SIS Conference at Fitzwilliam College, Cambridge, might give even the most hardened Darwinist pause for thought. It includes Professor Mark Bailey of Armagh Observatory, Dr. Mike Baillie of Queen's University Belfast, Dr. Victor Clube of Oxford University, Professor Gunnar Heinsohn of Bremen University, Dr. W. B. Masse of the University of Hawaii, Professor W. Mullen of Bard College, Professor David Pankenier of Lehigh University, Dr. Benny Pelser of Liverpool's John Moores University, and Professor Irving Wolfe of the University of Montreal. All are scheduled to speak on some aspect of geological catastrophes in the Bronze Age, counter to the prevailing trend of uniformitarian belief in gradualist geology.

Darwinism also remains a hot topic for discussion on the Internet where there are many news groups and conferences devoted to debating issues such as dating techniques and speciation and which resound daily with clashes between Darwinists and critics of all kinds.

One group of Darwinist vigilantes who are found regularly on the Internet are referred to and, indeed, proudly refer to themselves, in Internet jargon, as "howler monkeys." Readers will recall that howler monkeys gather in groups; have very loud voices that can carry as much as two or three miles; and enforce the boundaries of their territory by engaging in shouting matches with their enemies. Howlers also drive away their enemies by hurling handfuls of their own excrement at them.

The effects of the howler monkeys of the Internet are profoundly damaging to academic freedom of expression, whoever their current victim happens to be. In 1996, for instance, Dr. Peter Nyikos, professor of mathematics at the University of South Carolina, was rash enough to post some highly perceptive observations regarding the attempts by "cladists" to draw up family trees of ancestors and descendants along Darwinian lines. Nyikos, who is not a creationist, infuriated Internet Darwinists by pointing out that devotees of cladistics actually use a language with which creationists should be quite comfortable.

Despite his academic standing, Nyikos was not even accorded the civility of a hearing. He was immediately barraged with abuse and buried under tons of technical "objections" which kept him busy and unable to discuss publicly the flaws in Darwinism.

The fact that Nyikos is not a creationist but an evolutionist himself does not save him from such treatment. Indeed, Dr. Nyikos told me, "even fellow believers in evolution, like myself, get flamed without mercy if they aren't good 'team players' for the 'howler monkey' side."

Needless to say, if dissenting senior academics and scientists get this kind of treatment on the Internet, outsiders like myself and other nonacademic critics are routinely howled down without even a pretense of courtesy—an unexpected outcome of the information superhighway that many hoped would bring about global freedom of expression, led by the example of the academic community.

It would be encouraging to think that the forces of academic censorship and the suppression of dissent were a thing of the past in today's open, multi-media communications-linked world. Sadly, the malign influence of those who appoint themselves scientific vigilantes is becoming, if anything, even more widespread. Richard Dawkins, for instance, has now been appointed professor of the public understanding of science at Oxford University.

Dawkins has already shown the kind of methods he uses to foster the "public understanding of science" when he covertly campaigned to have my article for the *Times Higher Education Supplement* suppressed. It is depressing to find that a professor of the public understanding of science interprets his role as meaning he must supervise the information that the public and academic community are allowed to see and hear, and hence prevent them from gaining access to evidence that contradicts the accepted Darwinian doctrines.

How are the rest of us to understand academic behavior such as this? I believe that Darwinism has not only become transformed from scientific theory to scientific ideology, it has now become transformed from ideology to scientific urban myth, probably the most pervasive myth of the twentieth century. Darwinism the urban myth has become so powerful that it has dazzled the public and many scientists alike with its aura of unchallengable certainty and authority.

As this book has set out to show, Darwinism the urban myth has many faces. There is the myth of radiometric dating; the myth of uniformitarian geology; the myth of a gradualist fossil record; the myth of beneficial mutations; the myth of natural selection; the myth that evolution is blind; the myth of the beak of the finch; the myth of the biogenetic law; the myth of vestigial organs; the myth of homology; the myth of the "missing link."

Perhaps in one sense this transformation to mythic status gives some grounds for optimism. After all, science has demonstrated an enviable track record at eventually destroying its own myths, however long they have persisted and whatever the attempts by inquisitions and censors to maintain them.

Unfortunately science has also demonstrated a historical predilection for the comfort of such myths, as philosopher Paul Feyerabend points out:

> The stability achieved, the semblance of absolute truth is nothing but the result of an absolute conformism. For how can we possibly test, or improve upon, the truth of a theory if it is built in such a manner that any conceivable event can be described, and explained, in terms of its principles? The only way of investigating such all-embracing principles is to compare them with a different set of equally all-embracing principles—but this way has been excluded from the very beginning. The myth is therefore of no objective relevance, it continues to exist solely as the result of the effort of the community of believers and of their leaders, be these now priests or Nobel prize winners.[6]

CHAPTER 25

Old Theories Never Die

I F EVEN ONE HUNDREDTH PART OF THE EVIDENCE presented in this book is correct, then it will be obvious to any thinking person that there is a huge question mark hanging over the central issues of the life sciences.

What makes this state of affairs even more remarkable is that very few of the experiments described in this book could be called new or revelatory: on the contrary, their conclusions must be well known to anyone currently working in the earth sciences or life sciences in any of the world's universities.

Guy Berthault's discoveries on sedimentation have been widely published in the geological literature. The finding of Cairns and Hall that bacteria can mutate in a directed way were published in *Nature* nearly a decade ago. The conclusions of Zuckermann's studies that showed *Australopithecus* is merely an extinct ape were published more than forty years ago.

Yet these and hundreds of similar findings have been quietly forgotten about and continue to be ignored by almost all professionals in the field of evolutionary biology. Where you would expect to find penetrating questions being asked of neo-Darwinism, there is only an insistence on adhering to the received wisdom of reductionist science. Where you would expect vigorous public debate, there is only a nervous, artificial consensus among academics and a complete absence of dialogue in the press and on television.

The current mood in biology was summed up recently by maverick biologist Rupert Sheldrake as, "Rather like working in Russia under Brehznev. Many biologists have one set of beliefs at work, their official beliefs, and another set, their real beliefs, which they can speak openly about only among friends. They may treat living things as mechanical in the laboratory but when they go home they don't treat their families as inanimate machines."

In such an atmosphere and in the absence of any scientific or public debate, the only picture of evolution with which informed members of the public are familiar are the views of extremist reductionist writers who strive to turn the mountains of Darwinian improbability into molehills of scientific certainty.

Looked at in this light, it seems that one of the most important unanswered questions becomes: why should science resist any radical review of Darwinist ideas so fanatically?

I believe the answer to this question is that to any intelligent, educated, reasonable person, neo-Darwinism appears to be unassailable because it appears to be the *only reasonable theory available*. The only alternative appears to be either a religious explanation, as represented by the doctrine of creation, or half-baked speculations about aliens and quantum mechanics.

In this respect neo-Darwinism is seen by many of its adherents as the citadel of rationalism against the incursion of the barbarians of unscientific New-Age thinking. One unexpected result of this fanatical defense is that scientific rationalism, which used to be a badge of honor and a beacon of hope for the future, has sometimes become the white sheet and hood of bigoted closed-minded thinkers. And what I find particularly fascinating about this kind of thinking is that it is pretty nearly the exact opposite of the truth.

The attraction of scientific reductionism and the motive for wishing to banish metaphysical thinking is not difficult to understand. Science seems to have provided *reasonable* naturalistic explanations for many of the most important philosophical questions: How did life arise? What is mankind's position in the scheme of things? What holds together the physical fabric of the world and keeps the stars in their courses? These answers seem final, or close to final, after thousands of years of doubt. But it is not the finality of these explanations, or the quality of evidence that supports them that makes them so acceptable. The key word in this explanation of the causes of scientific reductionism is "reasonable."

To the scientists of the Babylonian civilization, it seemed reasonable to believe that the Earth was flat and was held up by elephants standing on a giant sea turtle—even though their astronomy was highly developed and they had observed the curvature of the Earth's shadow moving across the Moon during eclipses. They held this view because they could not imagine a plausible alternative theory. The idea of a flat Earth held up by elephants was the *most reasonable* explanation available. Flatness seemed to fit their everyday experience, and, although highly improbable, elephants were far less improbable than any conceivable alternative. Yet, because it was based on faulty evidence, it was actually only a superstitious belief. What appeared to be the most reasonable view was really completely unreasonable.

The flat-Earth theory was rejected by Greek scientists such as Pythagoras, Hipparchus, and Aristotle who observed that the Sun and Moon were spherical and reasoned that the Earth would be too.

Once the flat-Earth viewpoint was deprived of the appearance of being reasonable, its wildly improbable nature became obvious. Today it seems surprising to us that anyone could have believed in such a theory, however limited their scientific knowledge.

I believe that something very similar is true of parts of Western science today. It actually contains some wildly improbable theories—as improbable as elephants holding up the Earth. Yet these theories appear to represent a reasonable view because they offer a natural-sounding mechanistic explanation that seems to be consonant with common sense and our essentially limited experience and understanding of the world.

Whole areas of the Western scientific model fit into this category: theories that seem as solid as rock and, indeed, are the foundations of much of Western thinking. Yet, in reality, they are at best unsubstantiated and at worst no more than superstitions. Among these flat-Earth superstitions, Darwinism stands out as being central.

The primary message of this book is: the world is full of people who want you to believe in their "ism"—Darwinism, Freudianism,

Marxism, and the rest. Don't accept anything they say unless they can substantiate it with scientific evidence, however persuasive their arguments, and however authoritative their position. Insist on consulting the primary sources of evidence yourself and make up your own mind.

In one sense it is not difficult to understand, and even share, the concerns that Darwinists feel at what will follow if their theory is discredited and discarded. Our scientific knowledge is hard-won: the darkness of superstition and pseudoscience is a terrifying prospect. Science is right to be tenacious in defense of its territory.

Yet the greatest strength of science is its openness to debate. Science is strong because errors are exposed through the process of experiment and open argument and counter argument. Science does not flourish because vigilante scientists appoint themselves to guard the gates against heretics. If the heresy is true it will become accepted. If it is false, it will be shown to be false, by rational discourse.

It is not scientific debate that a civilized society has to fear, but scientific censorship.

Most scientists today earn their living from the public purse in one way or another. In effect we the community employ scientists to tackle the difficult task of explaining that which we do not understand. This is no easy job to be sure, and one in which success may depend as much on luck as it does on skill and judgment. Because it is a difficult job, a tacit understanding has arisen that it would be bad form or unseemly to criticize science or scientists seriously, as if they were a banker who added up sums wrongly or a grocer who forgot to deliver the sausages.

I reject this tacit consensus. I am a customer for the scientific service that we pay scientists to provide and I have a customer complaint: I am not satisfied with the answers they have provided on the mechanism of evolution and I want them to go back to their laboratories and investigate further.

I believe it is high time that consumerism finds a voice in the public sector and in the academic world as effectively as it has in industry and commerce. And I do not accept the convention that scientists may be criticized only by their peers.

Finally, I believe that science and reason—tempered by intuition—offer the only real hope of discovering answers to these baffling questions and I wholeheartedly support the Western scientific method of enquiry. I am, though, concerned that many people, including some scientists, pay lip service to this idea while thinking and acting like intellectual Stalinists.

There is a strong streak of intellectual arrogance and intellectual authoritarianism running through the history of Darwinism, from Thomas Huxley and Charles Darwin (both openly racist) through to Julian Huxley, one of the principal architects of the neo-Darwinist theory in the twentieth century, who publicly advocated that people who were genetically abnormal (such as those mentally and physically handicapped by heredity) should be sterilized to relieve society of having to care for their offspring.

This authoritarian streak is still present in some Darwinists today. It is seen in the outrage and indignation with which they greet any reasoned attempt to expose the theory to debate and to the light of real evidence. I believe this reaction is caused by the psychological phenomenon of cognitive dissonance described by Leon Festinger, referred to earlier, rather than by any malicious intention on their part. But I also believe that the effect is the same as if it were intentionally malicious and hence it should be resisted by all people who prize their independence of mind.

Darwinism has never had much appeal for science outside of the English-speaking world, and has never appealed much to the American public (although popular with the U.S. scientific establishment in the past). However, its ascendancy in science, in both Britain and America, has been waning for several decades as its grip has weakened in successive areas: geology; paleontology; embryology; comparative anatomy. Now even geneticists are beginning to have doubts. It is only in mainstream molecular biology and zoology that Darwinism retains serious enthusiastic supporters.

As growing numbers of scientists begin to drift away from neo-Darwinist ideas, the revision of Darwinism at the public level is long overdue, and is a process that I believe has already started.

What are the prospects that those scientists who are still true believers will come out of the bunker and engage in real debate over the scientific issues of neo-Darwinism? Historically, they are not encouraging. One of the twentieth century's most distinguished scientists and Nobel laureates, physicist Max Planck, observed that;

"A new scientific truth does not triumph by convincing its opponents and making them see the light, but rather because its opponents eventually die, and a new generation grows up that is familiar with it."

More simply: old theories never die, only their supporters. It may be another decade or two before a new generation grows up and restores intellectual rigor to the study of evolutionary biology. We must wait, and hope.

Notes

Preface

1. *The Times* 8.25, 1992.
2. *New Statesman* 8.28, 1992.
3. *The Independent* 9.13, 1992.
4. *Nature* 8.27, 1992.
5. *The Sunday Times* 8.23, 1992.

Chapter 2: Through the Looking Glass

1. Francis Crick, 1970.
2. Jacques Monod, 1972.
3. See for example A. G. Cairns-Smith, 1982.
4. F. M. Broadhurst, 1964.
5. See Chapter 24.

Chapter 3: A Matter of Conjecture

1. Murray Eden, 1967.
2. Francis Crick, 1981.
3. John Thackray, 1980.
4. E. Barghoorn, 1971.
5. H. D. Pflug and H. Jaeschke-Boyer, 1979.
6. Quoted in Taylor, 1983.
7. Manfred Schidlowski, 1988.
8. Gustaf Arrhenius, 1996.

9. A. G. Cairns-Smith, 1982.
10. Hubert Yockey, 1978, 1981, 1992.
11. Robert Sauer, 1989, 1990.
12. Van Eysinga, 1975.
13. John Thackray, 1980.
14. Gavin de Beer, 1970.
15. Harold L. Levin, 1978.

Chapter 4: The Key to the Past?

1. Ronald Millar, 1972
2. For example, D. N. Wadia, 1966 edn., on the Siwalik Hills beds.
3. W. F. Libby, 1955.
4. Richard E. Lingenfelter, 1963.
5. Hans Seuss, 1965.
6. V. R. Switzer, 1967.
7. Melvin Cook, October 1968.
8. Ibid.
9. R. W. Fairbridge, 1964, 1984.

Chapter 5: Rock of Ages

1. Melvin Cook, 1966.

2. John L. Mero, 1965.
3. Melvin Cook, 1966.
4. Brent Dalrymple, 1984.
5. Melvin Cook, 1957.
6. Brent Dalrymple, 1984.
7. James Lovelock, 1987 edn.
8. Melvin Cook, 1966.
9. C. S. Noble and J. J. Naughton, 1968.
10. G. J. Funkhouser and J. J. Naughton, 1968.
11. Brent Dalrymple, 1992.
12. F. J. Fitch and J. A. Miller, 1969.
13. Richard Leakey, 1973.
14. Fitch, Miller, and Hooker, 1976.
15. G. H. Curtis et al., 1975.
16. Fitch, Miller, and Hooker, 1976.
17. McDougall, 1981.
18. Ibid.

Chapter 6: Tales from Before the Flood

1. Leonard Woolley, 1950.
2. P. M. Hurley, 1968.
3. Melvin Cook, 1966.
4. Ibid.
5. Charles Hapgood and James Campbell, 1958.

Chapter 7: Fashioned from Clay

1. Stephen Jay Gould, 1990.
2. R. W. Gallois, 1965.
3. C. O. Dunbar and John Rogers, 1957.
4. John L. Mero, 1965.
5. V. I. Sozansky, March 1973.

6. Omer Roup, December 1970.
7. Guy Berthault, 1986, 1988.

Chapter 8: An Element of Unreality

1. S. E. Hollingsworth, 1962.
2. Melvin Cook, 1966.
3. F. M. Broadhurst, 1964.

Chapter 9: When Worlds Collide

1. I. Velikovsky, 1950.
2. Alfred De Grazia et al., 1966.
3. I. Velikovsky, 1955.
4. Harry S. Ladd, 1959.
5. C. O. Dunbar, 1960.
6. D. N. Wadia, 1966 edn.
7. Edwin Colbert, 1968.
8. Edwin Colbert, 1965.

Chapter 10: The Record of the Rocks

1. George Simpson, 1961 edn.
2. Ibid.
3. Garrett Hardin, 1961.
4. George Simpson, 1961 edn.
5. David Norman, 1985.
6. Sankar Chatterjee, 1991.
7. J. G. O. Smart et al., 1966
8. J. H. Callomon, in P. C. Sylvester-Bradley (Ed.), 1968.
9. Raymond C. Moore (Ed.), 1957.
10. Norman Macbeth, 1974.
11. Ernst Mayr, 1953.
12. See Adrian Desmond, 1975.
13. W. R. Thompson, 1959.
14. Keith Simpson, 1978.

Chapter 11: Survival of the Fittest

1. C. H. Waddington, 1960.
2. G. G. Simpson, 1967.
3. Julian Huxley, 1963.
4. G. G. Simpson, 1964.
5. Gavin de Beer, 1970.

Chapter 12: Green Mice and Blue Genes

1. Ernst Mayr, 1963.

Chapter 13: The Beak of the Finch

1. Jonathan Weiner, 1994.
2. Charles Darwin, 1987 edn.
3. Jonathan Weiner, 1994.
4. Ibid.
5. Theodosius Dobzhansky, 1937 edn.
6. Ernst Mayr, 1942.
7. Theodosius Dobzhansky, 1951 edn.
8. Ernst Mayr, 1970.

Chapter 14: Of Cabbages and Kings

1. Julian Huxley, 1963.
2. Jacques Monod, 1970.
3. Entry on Human Genetics in *Encyclopaedia Britannica*, 1984.
4. Richard Dawkins, 1986.
5. Ibid.
6. Francis Crick, 1981.
7. David Norman, 1985.

Chapter 15: The Ghost in the Machine

1. Richard Dawkins, 1986.

2. Andrew Scott, 1988.
3. Francis Crick, 1966.
4. Ibid.
5. William Shrive, 1982.

Chapter 16: Pandora's Box

1. Michael Denton, 1985
2. Gavin de Beer, 1971.
3. Ibid.
4. Ibid.
5. Ernst Mayr, 1970.
6. M. D. Dayhoff, 1972.
7. A. Ferguson, 1980.

Chapter 17: Paradigm Lost

1. S. R. Scadding, May 1981.
2. George Simpson, 1961 edn.
3. A. J. White, 1989.
4. Gavin De Beer, 1964.
5. Ernst Mayr. 1960.
6. George Simpson, 1967.
7. George Simpson, Pittendrigh, and Tiffany, 1958.
8. Arthur Koestler, 1967.

Chapter 18: Down from the Trees

1. Ernst Haeckel, 1899.
2. A. J. White, 1989.
3. A. J. Kelso, 1974.
4. Richard E. Leakey, 1981.
5. Solly Zuckermann, in Huxley, Hardy, and Ford, 1954.
6. Charles Oxnard, 1984.
7. Philip Tobias, 1967.
8. A. J. White, 1989.
9. See John Reader, 1981.
10. Jack Stern and Randall Susman, 1983.

Chapter 19: Hopeful Monsters

1. Alan Durrant, 1958, 1962.
2. J. Hill, 1965.
3. C. A. Cullis, 1977.
4. Francis Crick, 1970.
5. Howard Temin, 1976.
6. Ted Steele, 1979.
7. Reg Gorczinski and Ted Steele, 1977.
8. Barry Hall, September 1990.
9. Richard Goldschmidt, 1940.
10. Stephen Jay Gould, 1977.
11. Rupert Sheldrake, 1983.
12. Gerald Geison, 1995.
13. Wilhelm Reich, 1948, 1979.
14. Robert Dew, 1989.
15. Fred Hoyle and Chandra Wickramasinghe, 1981.
16. Fred Hoyle, 1983.
17. Francis Crick, 1981.

Chapter 20: The Facts of Life

1. A. A. Moscona, 1959.
2. T. Humphreys, 1963.
3. B. I. Balinsky, 1951.
4. D. J. Pritchard, 1990.
5. C. H. Waddington, in John Lewis (Ed.), 1974.
6. Wayne Itano et al., 1990.
7. David Bohm, 1980.
8. Hans Driesch, Presidential Address to the Society for Psychical Research, 1926.

9. Alister Hardy, 1949, 1950, 1953.

Chapter 21: The Evolution of Evolution

1. George Simpson, 1961 edn.
2. David Raup, 1991.
3. Karl Marx, 1961 edn.

Chapter 22: On Being Thick Skinned

1. Garret Hardin, 1960.
2. Gavin De Beer, 1964.
3. M. and A. Wolsky, 1976.
4. E. Denton, 1971.
5. Richard Goldschmidt, 1940.

Chapter 23: The Fish That Walked

1. J. L. B. Smith, 1940.

Chapter 24: Angels Versus Apes

1. Warwick Collins, 1994. Unpublished letter to *Nature*.
2. Brent Dalrymple, 1992.
3. Jerry Bergman, 1996.
4. *Mysterious Origins of Man*, NBC TV broadcasts, 1996.
5. D. J. Leveson and D. E. Seidemann, 1996.
6. Paul Feyerabend, 1965.

Bibliography

Arrhenius, Gustaf. 1996. *Evidence for life on Earth before 3800 Million years ago* in *Nature* 384: 55 (7 November 1996).

Bakker, Robert T. 1971. *Dinosaur physiology and the origin of mammals* in *Evolution* 25: 636–658.

Balinski, B. I. 1951. *On the eye cup-lens correlation in some South African amphibians* in *Experimenta* 7: 180–181.

Barghoorn, E. 1971. *The oldest fossils* in *Scientific American* 224(5): 30.

Bergman, Jerry. 1996. *Censorship of Information on Origins* in *Creation Ex Nihilo Technical Journal* 10: 3.

Berthault, Guy. 1986. *Comptes-Rendus Academie des Science II* 303(17): 1569–1574 (3 December 1986).

Berthault, Guy. 1988. *Comptes-Rendus Academie des Science II* 306(11): 717–724 (16 February 1988).

Black, M. 1953. *The constitution of chalk.* Report of lecture in *Proceedings of the Geological Society* No. 1499: lxxxi–vi.

Bohm, David. 1980. *Wholeness and the Implicate Order.* Routledge &Kegan Paul, London.

Briggs, D., and Walters, S. M. 1984 edn. *Plant Variation and Evolution.* Cambridge University Press.

Broadhurst, F. M. 1964. *Some aspects of the palaeoecology on non-marine faunas and rates of sedimentation in the Lancashire coal measures* in *American Journal of Science* 262: 865.

Buckland, William. 1823. *Reliquiae Diluvianae.* John Murray, London.

Buckland, William. 1836. *Geology and Mineralogy Considered with Reference to Natural Theology.* Pickering, London.

Cairns, J., Overbaugh, J., and Miller, S., 1988. *The origin of mutants* in *Nature* 335: 142–145.

Cairns-Smith, A. G. 1982. *Genetic Takeover*. Cambridge University Press.

Callomon, J. H. 1968. In P. C. Sylvester-Bradley (ed.). *The Geology of the East Midlands*. Leicester University Press.

Cave, A. J. E., and Strauss, W. L. 1957. *Pathology and posture of Neanderthal man* in *Quarterly Review of Biology* 32: 348–363.

Chatterjee, Sankar. July 1991. *Cranial anatomy and relationships of a new Triassic bird from Texas* in *Philosophical Transactions of the Royal Society London B* 332: 277-342.

Colbert, Edwin. 1965. *The Age of Reptiles*. W. W. Norton, New York.

Colbert, Edwin. 1968. *Men and Dinosaurs*. E. P. Dutton, New York.

Conklin, Edwin G. 1943. *Man Real and Ideal*. Scribners, New York.

Cook, Melvin A. January 1957. *Where is the Earth's radiogenic helium?* in *Nature* 179: 213.

Cook, Melvin A. 1966. *Prehistory and Earth Models*. Max Parrish, London.

Cook, Melvin A. Oct. 1968. *Do radiological clocks need repair?* in *Creation Research Society Quarterly* 5: 70.

Cox, L. R. (ed.). 1967 edn. *British Mesozoic Fossils*. British Museum of Natural History, London.

Crick, Francis. 1966. *Of Molecules and Men*. University of Washington Press, Seattle.

Crick, Francis. 1970. In *Nature* 227: 561.

Crick, Francis. 1981. *Life Itself*. Macdonald, London.

Crick, Francis, and Orgel, Leslie. 1973. *Directed panspermia* in *Icarus* 19: 341.

Cullis, C. A. 1977. *Molecular aspects of the environmental induction of heritable changes in Flax* in *Heredity* 38: 129–154.

Curtis, G. H. J., Drake, T., Cerling, and Hampel. 1975. *Age of KBS Tuff in Koobi Fora Formation, East Rudolf, Kenya* in *Nature* 258: 395–98 (4 December 1975).

Dart, Raymond. 1925. *Australopithecus africanus: The Man-Ape of South Africa* in *Nature* 115: 195–199.

Darwin, Charles. 1901 edn. *The Descent of Man*. John Murray, London.

Darwin, Charles. 1902 edn. *The Origin of Species*. John Murray, London.

Dawkins, Richard. 1986. *The Blind Watchmaker*. Longmans, London.

Dayhoff, M. D. 1972. *Atlas of Protein Sequence and Structure*. National Biomedical Research Foundation, Silver Spring, Maryland, Vol. 5 Matrix 1: D–8.

De Beer, Gavin. 1964. *Atlas of Evolution*. Nelson, London.

De Beer, Gavin. 1970. *A Handbook of Evolution*. British Museum of Natural History, London.

De Beer, Gavin. 1971. *Homology: An Unsolved Problem*. Oxford University Press, London.

De Grazia, Alfred, Juergens, R. E., and Stecchini, L. C. 1966. *The Velikovsky Affair.* Sidgewick & Jackson, London.

De Terra, Helmut, and Paterson, T. T. 1939. *Studies on the Ice Age in India.* Harvard University Press.

Denton, E. 1971. *Reflectors and fishes* in *Scientific American* 224(1): 64.

Denton, Michael. 1985. *Evolution: A Theory in Crisis.* Burnett Books, London.

Desmond, Adrian. 1975. *The Hot Blooded Dinosaurs.* Blond & Briggs, London.

Dew, Robert. 1987. *An Air Germ Experiment* in *Annals of the Institute for Orgonomic Science* 4: 15–43 (September 1987).

Dobzhansky, Theodosius. 1937. *Genetics and the Origin of Species.* Columbia University Press, New York.

Dobzhansky, Theodosius. 1951. *Genetics and the Origin of Species* (3rd edition). Columbia University Press, New York.

Dobzhansky, Theodosius. 1984 edn. *Heredity* in *Encyclopaedia Britannica,* 15th edition.

Driesch, Hans. 1908. *The Science and Philosophy of the Organism.* London.

Driesch, Hans. 1925. *The Philosophy of Vitalism.* London.

Dunbar, C. O., 1960 edn. *Historical Geology.* John Wiley, New York.

Dunbar, C. O., and Rogers, John. 1957. *Principles of Stratigraphy.* John Wiley, New York.

Durrant, Alan. 1958. *Environmental conditioning of flax* in *Nature* 81: 928–929.

Durrant, Alan. 1962. *The environmental induction of heritable changes in Linum* in *Heredity* 17: 27–61.

Eden, Murray. 1967. *The inadequacy of neo-Darwinian Evolution as a scientific theory.* Massachusetts Institute of Technology conference paper.

Einstein, Albert, Podolsky, Boris, and Rosen, Nathan. 1935. *Can quantum-mechanical description of physical reality be considered complete?* in *Physical Review* 47: 777.

Eldredge, Niles, and Gould, Stephen Jay. 1960. *Phylogenetic Patterns and the Evolutionary Process.* Columbia University Press.

Fairbridge, R. W. 1984 edn. *Holocene* in *Encyclopaedia Britannica,* 15th edition.

Fairbridge, R. W. 1964. *African Ice-Age aridity* in A. E. M. Nairn (ed.). *Problems in Palaeoclimatology.*

Faul, Henry. 1959. *Doubts of the Palaeozoic time scale* in *Journal of Geophysical Research* 64: 1102.

Faul, Henry. 1960. *Geologic time scale* in *Bulletin of the Geological Society of America* 71: 637–644.

Faul, Henry. 1978. *A history of geologic time* in *American Scientist* 66: 159–165.

Ferguson, A. 1980. *Biochemical Systematics and Evolution*. Blackie, Glasgow.

Feyerabend, Paul. 1965. In R. G. Colodny (ed.). *Problems of Empiricism* in *Beyond the Edge of Certainty*. pp. 145–260.

Fitch, F. J., and Miller, J. A. 1970. *Radioisotopic Age Determinations of Lake Rudolph Artifact Sites* in *Nature* 242: 447 (18 April 1970).

Fitch, F. J., Findlater, I. C., Watkins, R. T., and Miller, J. A. 1974. *Dating of a rock succession containing fossil hominids at East Rudolph, Kenya* in *Nature* 251: 214 (20 September 1974).

Fitch, F. J., Hooker, P. J., and Miller, J. A. 1970. *40ar/39ar dating of the KBS tuff in Koobi Fora formation, East Rudolph, Kenya* in *Nature* 263: 740–42 (28 October 1976)

Fisher, Ronald. 1930. *The Genetical Theory of Natural Selection*. Oxford University Press.

Freedman, S., and Clauser, J. 1972. *Experimental test of local hidden variables theories* in *Physical Review Letters* 28: 938.

Funkhouser, J. G., and Naughton, J. J. July 1968. *Journal of Geophysical Research* 73: 4606.

Gallois, R. W. 1965. *The Wealden District*. Institute of Geological Sciences, London.

Geison, Gerald. 1996. *The Notebooks of Louis Pasteur*. Princeton University Press.

Goldschmidt, Richard. 1940. *The Material Basis of Evolution*. Yale University Press.

Gorczynski, Reg, and Steele, Edward. 1977. *Inheritance of acquired immunological tolerance to foreign histocompatibility antigens*, in *Proceedings of the National Academy of Sciences of the USA*, 2871.

Gould, Stephen Jay. 1977. *Ontogeny and Phylogeny*. Harvard University Press.

Gould, Stephen Jay. 1990. *Wonderful life: The Burgess Shales and the Nature of History*. Hutchinson Radius, London.

Haekel, Ernst. 1866. *The General Morphology of Organisms*. London.

Haekel, Ernst. 1876 edn. *The History of Creation*. London.

Haekel, Ernst. 1899. *The Last Link*. London.

Hall, Barry G. Sept. 1990. *Spontaneous point mutations that occur more often when advantageous than when neutral* in *Genetics* 126: 5–16.

Hapgood, Charles, and Campbell, James. 1958. *Earth's Shifting Crust*. Museum Press, London.

Hardin, Garrett. 1961. *Nature and Man's Fate*. Mentor, New York.

Hardy, Alister. May/June 1950. In *Journal of the Society for Psychical Research*. London.

Hardy, Alister. 1953. In *Proceedings of the Society for Psychical Research*, Vol. 50. London.

Hardy, Alister. 1965. *The Living Stream*. Collins, London

Heilman, Gerhard. 1926. *The Origin of Birds*. Witherby, London.

Hill, J. 1965. *Environmental induction of heritable changes in Nicotiana rustica* in *Nature* 207: 732–734.

Hollingsworth, S. E. 1962. *The climatic factor in the geological record* in *Quarterly Journal, Geological Society of London* 118: 13.

Holmes, Arthur. 1947. *The construction of a geological time scale* in *Transactions of the Geological Society of Glasgow* 21: 117–152.

Holmes, Arthur. 1960. *A revised geological time scale* in *Edinburgh Geological Society Transactions* 17(Part 3): 183–216.

Holmes, Arthur. 1965. *Principles of Physical Geology*. Ronald Press, New York.

Hoyle, Fred, and Wickramasinghe, C. 1978. *Lifecloud*. J. M. Dent, London.

Hoyle, Fred. 1983. *The Intelligent Universe*. Michael Joseph, London.

Humphreys, T. 1963. *Chemical dissolution and in vitro reconstruction of sponge cell adhesions* in *Developmental Biology* 8: 27–47.

Hurley, P. M. 1968. *The confirmation of continental drift* in *Scientific American* 218(4): 52.

Hutton, James. 1899 edn. In Archibald Geikie (ed.). *Theory of the Earth with Proofs and Illustrations*. Geological Society, London.

Huxley, Julian S. 1955. *Evolution and Genetics* in J. R. Newman (ed.). *What Is Science?* Simon & Schuster, New York.

Huxley, Julian S. 1963 edn. *Evolution—The Modern Synthesis*. Allen & Unwin, London.

Huxley, Julian S., Hardy, A. C., and Ford, E. B. (eds.). 1954. *Evolution as a Process*. Allen & Unwin, London.

Itano, Wayne, Heinzen, D. J., Bollinger, J. J., and Wineland, D. J. *Quantum Zeno effect* in *Physical Review A* 41(5): 2295–2300.

Kelso, A. J. 1974. *Physical Anthropology* J. P Lipincott, New York.

Kieth, M. S., and Anderson, G. M. Aug. 1963. *Radiocarbon dating: fictitious results with mollusk shells* in *Science*, August 16: 634.

Koestler, Arthur. 1967. *The Ghost in the Machine*. Hutchinson, London.

Koestler, Arthur. 1978. *The Case of the Midwife Toad*. Hutchinson, London.

Kuhn, Thomas. 1962. *The Structure of Scientific Revolutions*. Chicago University Press.

Ladd, Harry S. 1959. *Ecology, Paleontology and Stratigraphy* in *Science* 129: 72.

Lamarck, Jean Baptiste de. 1809. *Philosophie Zoologique*. Paris.

Leakey, Richard E. 1973. *Evidence for an advanced Plio–Pleistocene Hominid from East Rudolph, Kenya* in *Nature* 242: 447 (13 April 1973).

Leakey, Richard E. 1981. *The Making of Mankind*. Michael Joseph, London.

Levin, Harold L. 1978. *The Earth Through Time*. W. B. Saunders, Philadelphia.

Levison, D. J., and Seidemann, D. E. 1996. *Richard Milton—A non-religious creationist ally* in *Journal of Geoscience Education* 44: 428–438.

Lewis, John (ed.). 1974. *Beyond Chance and Necessity*. Garnstone Press, London.

Libby, Willard F. 1955 edn. *Radiocarbon Dating*. Chicago University Press.

Lingenfelter, Richard E. Feb. 1963. *Production of C-14 by cosmic ray neutrons* in *Review of Geophysics* 1: 51.

Lovelock, James E. 1987 edn. *Gaia: A New Look at Life on Earth*. Oxford University Press.

Lowe, D. 1980. *Stromatolites 3400 Myr-old from the Archean of Western Australia* in *Nature* 284: 441.

Lyell, Charles. 1833 (edn.) *Principles of Geology*. John Murray, London.

Macbeth, Norman. 1974 edn. *Darwin Retried*. Garnstone Press, London.

Mantell, Gideon. 1822. *The Fossils of the South Downs, or Illustrations of the Geology of Sussex*. Relfe, London.

Mantell, Gideon. 1825. *On the teeth of the Iguanodon* in *Philosophical Transactions of the Royal Society* 115: 179–186.

Marx, Karl. 1961 edn. *Capital*. Foreign Languages Publishing House, Moscow.

Mayr, Ernst. 1953. *Methods and Principles of Systematic Zoology*. McGraw Hill, New York.

Mayr, Ernst. 1963. *Animal Species and Evolution*. Harvard University Press.

Mayr, Ernst. 1960. *The emergence of evolutionary novelties* in *Tax* 1: 349–80.

Mayr, Ernst. 1964 edn. *Systematics and the Origin of Species*. Dover, New York.

Mayr, Ernst. 1970. *Population, Species and Evolution*. Harvard University Press.

McDougall, I., Maier, R., Sutherland-Hawkes, P., and Gleadow, A. J. W. 1980. *K-Ar age estimate for the KBS Tuff. East Turkana, Kenya* in *Nature* 284: 230–32 (20 March 1980).

McDougall, I. 1981. *40Ar/39Ar age spectra from the KBS Tuff, Koobi Fora formation* in *Nature* 294: 123–24 (12 November 1981).

Mero, John L. 1965. *The Mineral Resources of the Sea*. Elsevier, London.

Millar, Ronald, 1972. *The Piltdown Men*. Gollancz, London.

Miller, Hugh. 1869 edn. *The Old Red Sandstone.* W. P. Nimmo, Edinburgh.

Miller, Stanley L. 1953. *A production of amino acids under possible primitive Earth conditions* in *Science* 117: 528–529.

Milton, Richard. 1996. *Alternative Science: Challenging the Myths of the Scientific Establishment.* Park Street Press, Vermont.

Monod, Jacques, 1972 edn. *Chance and Necessity.* William Collins, Glasgow.

Moore, Raymond C. (ed.). 1957. *Treatise on Invertebrate Paleontology. Part L—Mollusca.* Geological Society of America and Kansas University Press.

Moorehead, Alan. 1969. *Darwin and the Beagle.* Hamish Hamilton, London.

Morris, Henry M. (ed.). 1985 edn. *Scientific Creationism.* Institute of Creation Research, San Diego.

Moscona, A. A. 1959. *Tissues from dissociated cells* in *Scientific American* 200(5): 132.

Noble, C. S., and Naughton, J. J. Oct. 1968. *Deep ocean basalts: Inert gas content and uncertainties in age dating* in *Science* 162: 265.

Norman, David. 1985. *Encyclopaedia of Dinosaurs.* Salamander Books, London.

Opdyke, N. D., and Runcorn, S. K. 1956. *New evidence for the reversal of the geomagnetic field near the Plio-Pleistocene boundary* in *Science* 123 (3208).

Opdyke, N. D. 1968. *Palaeomagnetism of Oceanic Cores* in R. A. Phinney (ed.). *History of the Earth's Crust.*

Oxnard, Charles. 1984. *The Order of Man.* Yale University Press.

Paley, William. 1828. *Natural Theology.* London.

Pflug, H. D., and Jaeschke-Boyer, H. 1979. *Combined structural and chemical analysis of 3800 Myr-old microfossils* in *Nature* 280: 483.

Phinney, R. A. 1968. *History of the Earth's crust.* Princeton University Press.

Pritchard, D. J. 1990. *The missing chapter in evolution theory* in *Biology* 37 (5): 149–152.

Raup, David. 1991. *Extinction: Bad genes or bad luck?* in *New Scientist* 14: 46–49 (September 1991).

Reader, John. 1981. *Missing Links.* Collins. London.

Reich, Wilhelm. 1979. *The Bion Experiments on the Origin of Life.* Farrar, Strauss, & Giroux, New York.

Romer, A. S. 1966 edn. *Vertebrate Palaeontology.* University of Chicago Press, Chicago/London.

Romer, A. S. 1968. *The Process of Life.* Weidenfeld & Nicholson. London.

Roup, Omer B. Dec. 1970. *Brine mixing: An additional mechanism for formation of basin evaporites* in *Bulletin, American Association of Petroleum Geologists* 54: 2258.

Sauer, R. T., Bowie, J. U., Olson, J. F. R., and Lim, W. A. 1989. *Proceedings of the National Academy of Sciences* 86: 2152–2156.

Scadding, S. R. May 1981. *Do vestigial organs provide evidence for evolution* in *Evolutionary Theory* 5: 173–176.

Scheuchzer, Johann. 1726. *Homo Diluvii Testis*. Burkli, Zurich.

Schidlowski Manfred. 1988. *A 3800-million year isotopic record of life from carbon in sedimentary rocks* in *Nature* 333: 313–318.

Schrodinger, Erwin. 1955 edn. *What is Life?* Cambridge University Press.

Scott, Andrew. 1988. *Vital Principles*. Blackwell, Oxford.

Sheldrake, Rupert. 1988 edn. *A New Science of Life*. Paladin, London.

Sheldrake, Rupert. 1994. *Seven Experiments That Could Change The World*. Fourth Estate, London.

Shrive, William. 1982. *Enzymes* in *Encyclopaedia of Science and Technology*.

Simpson, Keith. 1978. *Forty Years of Murder*. Harrap & Co., London.

Simpson, George G. 1944. *Tempo and Mode in Evolution*. Columbia University Press.

Simpson, George G. 1961 edn. *Horses*. Oxford University Press.

Simpson, George G., Pittendrigh, C. S., and Tiffany, L. H. 1958. *Life: An Introduction to Biology*. Routledge & Keegan Paul, London.

Simpson, George G. 1967 edn. *The Meaning of Evolution*. Yale University Press.

Simpson, George G. 1969. *Biology and Man*. Harcourt Brace.

Smart, J. G. O., et al. 1966. *Geology of the country around Canterbury and Folkestone*, HMSO for Geological Survey of Great Britain.

Smith, Andrew B. 1987. *Fossils of the Chalk*. The Palaeontological Association, London.

Smith, J. L. B. 1940. *A Living Coelacanthid Fish from South Africa* in *Transactions of the Royal Society of Africa* 28:1–106.

Sozansky, V. I. March 1973. *Origin of salt deposits in deep-water basins of Atlantic ocean* in *Bulletin, American Association of Petroleum Geologists* 57: 590.

Spath, L. F. 1923–1943. *Ammonoidea of the Gault*. The Palaeontographical Society, London.

Steele, Edward (ed.). 1979. *Genetic Selection and Adaptive Evolution*. Williams & Wallace, Toronto.

Stern, J. T., and Susman, R. 1983. *Australopithecus afarensis* in *American Journal of Physical Anthropology* 60: 279.

Suess, Hans E. December 1965. *Secular variations in the Cosmic-ray produced Carbon-14 in the atmosphere and their interpretations* in *Journal of Geophysical Research* 70: 5947.

Switzer, V. R. Aug. 1967. *Radioactive dating and low level counting* in *Science* 157: 726.

Taylor, Gordon Rattray. 1983. *The Great Evolution Mystery.* Secker & Warburg, London.

Temin, Howard M. 1976. *The DNA provines hypothesis* in *Science* 192: 1075.

Thackray, John. 1980. *The Age of the Earth.* Institute of Geological Sciences, London.

Thompson, W. R. 1956. Introduction to *The Origin of Species.* Everyman Library, Dutton, London.

Thomson, William (Lord Kelvin). 1862. *On the secular cooling of the earth* in *Royal Society of Edinburgh Transactions* 23(1): 157–169.

Tobias, P. V. 1967. *Olduvai Gorge 2: The cranium and maxillary dentition of Australopithecus (Zinjanthropus) boisei.* Cambridge.

Van Eysinga, F. W. B. 1975 edn. *Geological Time Table.* Elsevier Scientific, Amsterdam.

Velikovsky, Immanuel. 1973 edn. *Earth in Upheaval.* Sphere, London.

Velikovsky, Immanuel. 1973 edn. *Worlds in Collision.* Sphere, London.

Vine, F. J. 1968. *Magnetic Anomalies Associated with Mid-Ocean Ridges* in R. A. Phinney (ed.). *History of the Earth's Crust.*

Waddington, C. H. 1960. *Evolutionary adaptation* in *Tax* 1: 381–402.

Waddington, C. H. 1974. *How much is evolution affected by chance and necessity?* in John Lewis (ed.). *Beyond Chance and Necessity.* Garnstone Press, London.

Wadia, D. N. 1966 edn. *Geology of India.* Macmillan, London.

Wegener, Alfred. 1924. *The Origin of Continents and Oceans.* Methuen, London.

Weidersheim, Ernst. 1895. *The Structure of Man.*

Weisman, August. 1893. *The Germ Plasm: A Theory of Heredity.* W. Scott, London.

White, A. J. 1989. *Wonderfully Made.* Evangelical Press, Darlington.

Wolksy, M., and A. 1976. *The Mechanism of Evolution: A New Look at Old Ideas.* Karger, Basel.

Woolley, Leonard. 1950 edn. *Digging up the Past.* Penguin Books, London.

Yockey, Hubert P. 1978. *On the information content of cytochrome C and the calculation of probability of spontaneous biogenesis by information theory* in *Journal of Theoretical Biology* 67.

Yockey, Hubert P. 1981. *Self organization origin of life scenarios and information theory* in *Journal of Theoretical Biology* 91.

Yockey, Hubert P. 1992. *Information Theory and Molecular Biology,* Cambridge University Press.

Zuckermann, Solly. 1954. *Correlation of changes in the evolution of the higher primates* in Huxley, Hardy, and Ford (eds.). *Evolution as a Process.*

Zukav, Gary. 1979. *The Dancing Wu Li Masters.* Hutchinson London.

Other sources

The Mysterious Origins of Man, produced by Bill Cote, Carol Cote, and John Cheshire, broadcast by NBC Network TV in February and May 1996. BC Video, New York.

Comparative Evolution, by Warwick Collins. Unpublished article submitted to *Nature* (7 January 1994).

General Index

People Index

Species Index

305